Scrapy网络爬虫实战

东郭大猫 著

清華大學出版社
北京

内 容 简 介

随着大数据分析、大数据计算火热兴起，越来越多的企业发布了数据分析岗位，而数据分析的基础则是海量的数据。Python 中的 Scrapy 框架就是为了抓取数据而设计的。本书是一本 Scrapy 爬虫框架零基础起步的实战图书。

本书共分 11 章，第 1~2 章介绍 Python 环境的搭建、编辑器的使用、爬虫的一些基础知识（urllib、requests、Selenium、Xpath、CSS、正则表达式、BeautifulSoup 库）等。第 3~8 章主要介绍 Scrapy 框架的原理与使用。第 9~11 章主要介绍 Scrapy 的优化，包括内置服务、组件优化等，最后通过一个完整的大型示例对全书的知识点做了应用总结。

本书入门门槛低、浅显易懂，适合所有 Scrapy 爬虫和数据分析行业的入门读者学习，也适合高等院校和培训学校作为爬虫入门教材和训练手册。

图书在版编目（CIP）数据

Scrapy 网络爬虫实战/东郭大猫著. —北京：清华大学出版社，2019（2020.10重印）
ISBN 978-7-302-53620-8

Ⅰ. ①S… Ⅱ. ①东… Ⅲ. ①软件工具－程序设计 Ⅳ. ①TP311.561

中国版本图书馆 CIP 数据核字（2019）第 173907 号

责任编辑：夏毓彦
封面设计：王　翔
责任校对：闫秀华
责任印制：丛怀宇

出版发行：清华大学出版社
网　　址：http://www.tup.com.cn，http://www.wqbook.com
地　　址：北京清华大学学研大厦 A 座　　**邮　　编：**100084
社 总 机：010-62770175　　**邮　　购：**010-62786544
投稿与读者服务：010-62776969，c-service@tup.tsinghua.edu.cn
质 量 反 馈：010-62772015，zhiliang@tup.tsinghua.edu.cn

印 装 者：北京嘉实印刷有限公司
经　　销：全国新华书店
开　　本：190mm×260mm　　**印　　张：**15.75　　**字　　数：**403 千字
版　　次：2019 年 10 月第 1 版　　**印　　次：**2020 年 10 月第 2 次印刷
定　　价：59.00 元

产品编号：081510-01

前　言

读懂本书

还在复制粘贴找数据？

我想要这个网站上的数据，该怎么办？打开网站，复制，打开文本，粘贴……重复、重复、重复。

——费时、费力、错误多！

讲解晦涩难懂？

道理我都懂，可是要怎么做？这些数据我都想要，可是要怎么开始？本书不仅介绍 Scrapy 爬虫的原理，而且还给出实战案例让读者应用它们。

——爬虫的使用才是硬道理。

本书真的适合你吗？

本书帮你从零开始学习 Scrapy 爬虫技术，从基本的网络请求原理到抓取数据的保存，从单页面数据的下载到全站数据的爬取，从文本文档到数据库存储，本书介绍了实际使用中的各种基础知识。

——爬虫零基础？没关系，本书给出了从零开始学习的新手方案。

本书涉及的技术或框架

Python　　HTTP　　MySQL

Requests　　JSON　　MongoDB

BeautifulSoup　　XPATH　　Visual Studio

Selenium　　CSS　　Chrome 调试

本书涉及的示例和案例

抓取知乎热榜
名言网站抓取
博客园 Python 类文章抓取
深圳市社会保障局下载中心文件下载
链家数据保存至 MongoDB
豆瓣使用 Cookies 登录
抓取 cnBeta 科技类文章
伯乐在线订阅源数据抓取
伯乐在线最新文章抓取保存
起点小说网站小说封面抓取
豆瓣模拟提交表单登录
使用代理与统计链家小区信息
名言网站数据统计
IT 之家新闻抓取

猫眼电影 top100 抓取并发送邮件　　　　SegmentFault 全网用户信息抓取

本书特点

（1）本书不论是爬虫基础知识的介绍还是实例的开发，都是从实际应用的角度出发，精心选择典型的例子，讲解细致，分析透彻。

（2）深入浅出、轻松易学，以实例为主线，激发读者的学习兴趣，让读者能够快速学会 Scrapy 爬虫的实用技术。

（3）技术新颖、与时俱进，结合时下实用的技术，如 Requests、BeautifulSoup、Scrapy，使读者能够真正运用到实际工作中。

（4）贴近读者、贴近实际，大量成熟的第三方库和框架的使用和说明，帮助读者快速找到问题的最优解决方案，书中很多实例来自作者常用的数据源。

示例代码下载

本书示例代码请扫描二维码获得。如果下载有问题，请联系 booksaga@163.com，邮件主题为“Scrapy 网络爬虫实战”。

本书适用读者

Scrapy 网络爬虫初学者

从事 Web 网络数据分析的人员

从事数据存储的工作人员

高校与培训学校的师生

作　者

2019 年 5 月

目　录

第 1 章　Python 开发环境的搭建 …… 1
1.1　Python SDK 安装 …… 1
1.1.1　在 Windows 上安装 Python …… 1
1.1.2　在 Ubuntu 上安装 Python …… 2
1.2　安装开发工具 PyCharm 社区版 …… 3
1.3　安装开发工具 Visual Studio 社区版 …… 5
第 2 章　爬虫基础知识 …… 6
2.1　爬虫原理 …… 6
2.1.1　爬虫运行基本流程 …… 6
2.1.2　HTTP 请求过程 …… 8
2.2　网页分析方法 1：浏览器开发人员工具 …… 9
2.2.1　Elements 面板 …… 10
2.2.2　Network 面板 …… 11
2.3　网页分析方法 2：XPath 语法 …… 14
2.3.1　XPath 节点 …… 14
2.3.2　XPath 语法 …… 15
2.3.3　XPath 轴 …… 17
2.3.4　XPath 运算符 …… 19
2.4　网页分析方法 3：CSS 选择语法 …… 19
2.4.1　元素选择器 …… 20
2.4.2　类选择器 …… 21
2.4.3　ID 选择器 …… 21
2.4.4　属性选择器 …… 21
2.4.5　后代选择器 …… 21
2.4.6　子元素选择器 …… 22
2.4.7　相邻兄弟选择器 …… 22
2.5　网页分析方法 4：正则表达式 …… 22
2.5.1　提取指定字符 …… 23
2.5.2　预定义字符集 …… 23
2.5.3　数量限定 …… 23

2.5.4 分支匹配 ······ 24
2.5.5 分组 ······ 24
2.5.6 零宽断言 ······ 24
2.5.7 贪婪模式与非贪婪模式 ······ 25
2.5.8 Python 中的正则表达式 ······ 25
2.6 爬虫常用类库 1：Python 中的 HTTP 基本库 urllib ······ 30
2.6.1 发送请求 ······ 30
2.6.2 使用 Cookie ······ 31
2.7 爬虫常用类库 2：更人性化的第三方库 requests ······ 33
2.7.1 发送请求 ······ 34
2.7.2 请求头 ······ 35
2.7.3 响应内容 ······ 35
2.7.4 响应状态码 ······ 36
2.7.5 cookies 参数 ······ 37
2.7.6 重定向与请求历史 ······ 37
2.7.7 超时 ······ 38
2.7.8 设置代理 ······ 38
2.7.9 会话对象 ······ 38
2.8 爬虫常用类库 3：元素提取利器 BeautifulSoup ······ 39
2.8.1 安装 BeautifulSoup ······ 39
2.8.2 安装解析器 ······ 40
2.8.3 BeautifulSoup 使用方法 ······ 41
2.8.4 BeautifulSoup 对象 ······ 43
2.8.5 遍历文档树 ······ 47
2.8.6 搜索文档树 ······ 52
2.8.7 BeautifulSoup 中的 CSS 选择器 ······ 57
2.9 爬虫常用类库 4：Selenium 操纵浏览器 ······ 58
2.9.1 安装 Selenium ······ 59
2.9.2 Selenium 的基本使用方法 ······ 59
2.9.3 Selenium Webdriver 的原理 ······ 61
2.9.4 Selenium 中的元素定位方法 ······ 61
2.9.5 Selenium Webdriver 基本操作 ······ 63
2.9.6 Selenium 实战：抓取拉钩网招聘信息 ······ 64
2.10 爬虫常用类库 5：Scrapy 爬虫框架 ······ 67
2.10.1 安装 Scrapy ······ 67
2.10.2 Scrapy 简介 ······ 68

2.11 基本爬虫实战：抓取 cnBeta 网站科技类文章……69
2.11.1 URL 管理器……70
2.11.2 数据下载器……71
2.11.3 数据分析器……72
2.11.4 数据保存器……74
2.11.5 调度器……75
第 3 章 Scrapy 命令行与 Shell……78
3.1 Scrapy 命令行介绍……78
3.1.1 使用 startproject 创建项目……80
3.1.2 使用 genspider 创建爬虫……81
3.1.3 使用 crawl 启动爬虫……82
3.1.4 使用 list 查看爬虫……82
3.1.5 使用 fetch 获取数据……83
3.1.6 使用 runspider 运行爬虫……84
3.1.7 通过 view 使用浏览器打开 URL……85
3.1.8 使用 parse 测试爬虫……85
3.2 Scrapy Shell 命令行……85
3.2.1 Scrapy Shell 的用法……85
3.2.2 实战：解析名人名言网站……86
第 4 章 Scrapy 爬虫……89
4.1 编写爬虫……89
4.1.1 scrapy.Spider 爬虫基本类……89
4.1.2 start_requests()方法……90
4.1.3 parse(response)方法……91
4.1.4 Selector 选择器……91
4.2 通用爬虫……94
4.2.1 CrawlSpider……94
4.2.2 XMLFeedSpider……95
4.2.3 CSVFeedSpider……96
4.2.4 SitemapSpider……97
4.3 爬虫实战……98
4.3.1 实战 1：CrawlSpider 爬取名人名言……98
4.3.2 实战 2：XMLFeedSpider 爬取伯乐在线的 RSS……102
4.3.3 实战 3：CSVFeedSpider 提取 csv 文件数据……104
4.3.4 实战 4：SitemapSpider 爬取博客园文章……106

第 5 章 Scrapy 管道 109

5.1 管道简介 109
5.2 编写自定义管道 110
5.3 下载文件和图片 113
5.3.1 文件管道 114
5.3.2 图片管道 117
5.4 数据库存储 MySQL 121
5.4.1 在 Ubuntu 上安装 MySQL 121
5.4.2 在 Windows 上安装 MySQL 122
5.4.3 MySQL 基础 125
5.4.4 MySQL 基本操作 127
5.4.5 Python 操作 MySQL 129
5.5 数据库存储 MongoDB 131
5.5.1 在 Ubuntu 上安装 MongoDB 132
5.5.2 在 Windows 上安装 MongoDB 132
5.5.3 MongoDB 基础 135
5.5.4 MongoDB 基本操作 137
5.5.5 Python 操作 MongoDB 143
5.6 实战：爬取链家二手房信息并保存到数据库 144

第 6 章 Request 与 Response 157

6.1 Request 对象 157
6.1.1 Request 类详解 158
6.1.2 Request 回调函数与错误处理 160
6.2 Response 162
6.2.1 Response 类详解 162
6.2.2 Response 子类 163

第 7 章 Scrapy 中间件 165

7.1 编写自定义 Spider 中间件 165
7.1.1 激活中间件 165
7.1.2 编写 Spider 中间件 166
7.2 Spider 内置中间件 168
7.2.1 DepthMiddleware 爬取深度中间件 168
7.2.2 HttpErrorMiddleware 失败请求处理中间件 168
7.2.3 OffsiteMiddleware 过滤请求中间件 169
7.2.4 RefererMiddleware 参考位置中间件 169

7.2.5 UrlLengthMiddleware 网址长度限制中间件 ······170
7.3 编写自定义下载器中间件······170
7.3.1 激活中间件······170
7.3.2 编写下载器中间件······171
7.4 下载器内置中间件······173
7.4.1 CookiesMiddleware······173
7.4.2 HttpProxyMiddleware······174
7.5 实战：为爬虫添加中间件······174
第 8 章 Scrapy 配置与内置服务······178
8.1 Scrapy 配置简介······178
8.1.1 命令行选项（优先级最高）······178
8.1.2 每个爬虫内配置······179
8.1.3 项目设置模块······179
8.1.4 默认的命令行配置······181
8.1.5 默认全局配置（优先级最低）······182
8.2 日志······182
8.3 数据收集······184
8.4 发送邮件······187
8.4.1 简单例子······187
8.4.2 MailSender 类······187
8.4.3 在 settings.py 中对 Mail 进行设置······188
8.5 实战：抓取猫眼电影 TOP100 榜单数据······188
8.5.1 分析页面元素······189
8.5.2 创建项目······189
8.5.3 编写 items.py······190
8.5.4 编写管道 pipelines.py······190
8.5.5 编写爬虫文件 top100.py······191
第 9 章 模拟登录······194
9.1 模拟提交表单······194
9.2 用 Cookie 模拟登录状态······197
9.3 项目实战······198
9.3.1 实战 1：使用 FormRequest 模拟登录豆瓣······198
9.3.2 实战 2：使用 Cookie 登录······202

第10章 Scrapy爬虫优化 205
10.1 Scrapy+MongoDB实战：抓取并保存IT之家博客新闻 205
10.1.1 确定目标 205
10.1.2 创建项目 206
10.1.3 编写items.py文件 207
10.1.4 编写爬虫文件news.py 207
10.1.5 编写管道pipelines.py 209
10.1.6 编写settings.py 210
10.1.7 运行爬虫 211
10.2 用Benchmark进行本地环境评估 212
10.3 扩展爬虫 214
10.3.1 增大并发 214
10.3.2 关闭Cookie 214
10.3.3 关闭重试 214
10.3.4 减少下载超时时间 215
10.3.5 关闭重定向 215
10.3.6 AutoThrottle扩展 215
第11章 Scrapy项目实战：爬取某社区用户详情 217
11.1 项目分析 217
11.1.1 页面分析 217
11.1.2 抓取流程 221
11.2 创建爬虫 221
11.2.1 cookies收集器 222
11.2.2 Items类 225
11.2.3 Pipeline管道编写 226
11.2.4 Spider爬虫文件 227
11.2.5 Middlewars中间件编写 235

第 1 章

Python 开发环境的搭建

Scrapy 是使用 Python 编写的爬虫框架，在使用 Scrapy 之前，需要搭建开发环境。本章将为大家介绍 Python 的安装以及一些实用的编辑器的安装，以方便我们后续的开发工作。熟悉 Windows 系统的读者可以选择在 Windows 上搭建本书开发环境。

本章的主要知识点有：

- Python 安装
- PyCharm 编辑器安装
- Visual Studio 编辑器安装

1.1 Python SDK 安装

Python 是跨平台语言，可以运行在 Windows、Mac 及 Linux/UNIX 系统上，因此编写的代码在平台上没有运行的限制。目前，Python 有 Python 2 和 Python 3 两个版本，不幸的是两个版本很多地方不兼容。由于 Python 2 即将停止支持，而越来越多的库已经支持 Python 3，况且 Python 3 也提供了很多 Python 2 没有的新功能，因此本书使用 Python 3 搭建环境。

1.1.1 在 Windows 上安装 Python

首先从 Python 官网（https://www.python.org/downloads/）下载安装包，本书使用的是 3.6.3 版本，读者也可以下载更新的版本。下载后文件名为 python-3.6.3-amd64.exe，双击进行安装。

（1）在安装时先勾选 Add Python 3.6 to PATH 复选框，再选择 Customize installation 选项，如图 1.1 所示。

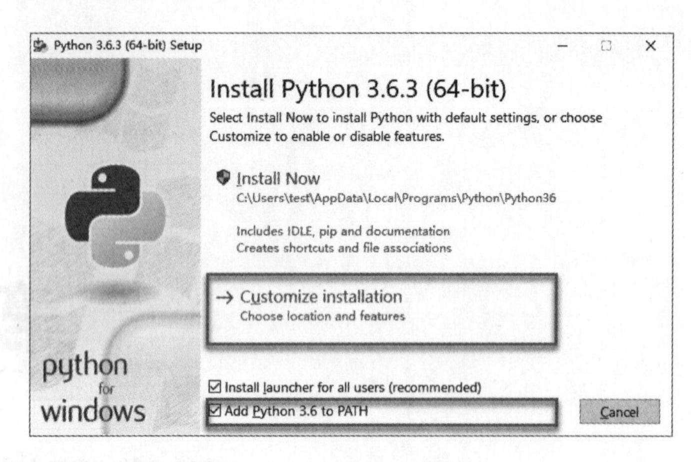

图 1.1　Python 3.6 安装首页

（2）务必选中 pip 复选框，单击 Next 按钮进行安装，如图 1.2 所示。

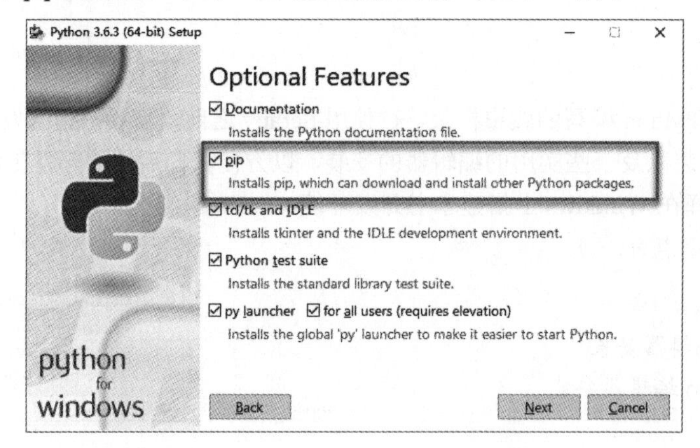

图 1.2　pip 选项

（3）选择安装路径，默认安装即可。打开命令提示窗口，输入 python，若出现 Python 版本号，则进入 Python 交互页面“>>>”，说明安装成功，如图 1.3 所示。

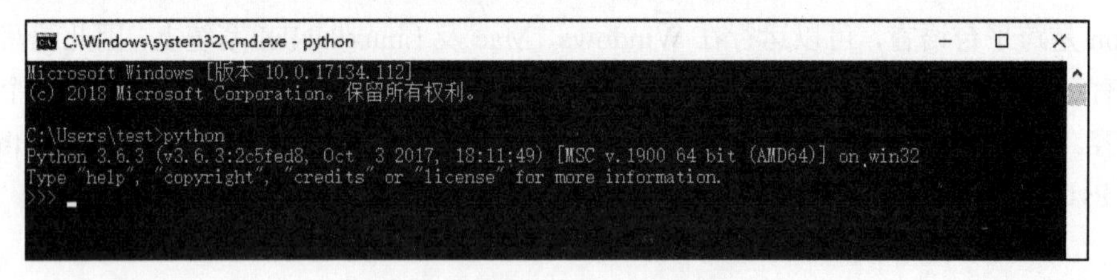

图 1.3　Python 3.6 安装检测

1.1.2　在 Ubuntu 上安装 Python

本书 Linux 发行版使用的是 Ubuntu 18.04，读者也可选择其他 Ubuntu 版本或者其他 Linux 发行版。Ubuntu 18.04 已经安装好了 Python 3，版本为 3.6.5。打开终端，输入 python3 命令，如图 1.4 所示。

```
scrapy@mypc: ~
文件(F) 编辑(E) 查看(V) 搜索(S) 终端(T) 帮助(H)
To run a command as administrator (user "root"), use "sudo <command>".
See "man sudo_root" for details.

scrapy@mypc:~$ python3
Python 3.6.5 (default, Apr  1 2018, 05:46:30)
[GCC 7.3.0] on linux
Type "help", "copyright", "credits" or "license" for more information.
>>>
```

图 1.4　Python 3.6 安装检测

这里我们需要手动安装 pip，使用命令 sudo apt install python3-pip 进行安装，安装完成之后，运行 pip3，结果如图 1.5 所示。

```
scrapy@mypc: ~
文件(F) 编辑(E) 查看(V) 搜索(S) 终端(T) 帮助(H)
scrapy@mypc:~$ pip3

Usage:
  pip <command> [options]

Commands:
  install                     Install packages.
  download                    Download packages.
  uninstall                   Uninstall packages.
  freeze                      Output installed packages in requirements format.
  list                        List installed packages.
  show                        Show information about installed packages.
  check                       Verify installed packages have compatible dependencies.
  search                      Search PyPI for packages.
  wheel                       Build wheels from your requirements.
  hash                        Compute hashes of package archives.
  completion                  A helper command used for command completion.
  help                        Show help for commands.

General Options:
  -h, --help                  Show help.
  --isolated                  Run pip in an isolated mode, ignoring
                              environment variables and user configuration.
  -v, --verbose               Give more output. Option is additive, and can be
                              used up to 3 times.
  -V, --version               Show version and exit.
  -q, --quiet                 Give less output. Option is additive, and can be
                              used up to 3 times (corresponding to WARNING,
                              ERROR, and CRITICAL logging levels).
  --log <path>                Path to a verbose appending log.
  --proxy <proxy>             Specify a proxy in the form
                              [user:passwd@]proxy.server:port.
  --retries <retries>         Maximum number of retries each connection should
                              attempt (default 5 times).
  --timeout <sec>             Set the socket timeout (default 15 seconds).
  --exists-action <action>    Default action when a path already exists:
                              (s)witch, (i)gnore, (w)ipe, (b)ackup, (a)bort.
```

图 1.5　pip 安装检测

1.2　安装开发工具 PyCharm 社区版

安装好 Python SDK 之后，我们需要一个方便的 IDE 来编写脚本。一个好的 IDE 能极大地提高工作效率，编者使用的是 PyCharm 这款编辑器。PyCharm 分为社区版和专业版，社区版为免费版本；专业版需付费，并且提供了更多的功能。针对爬虫开发来说，社区版已经足够使用了。

（1）进入 PyCharm 下载页面，以安装 Windows 版为例，下载社区版安装包，如图 1.6 所示。

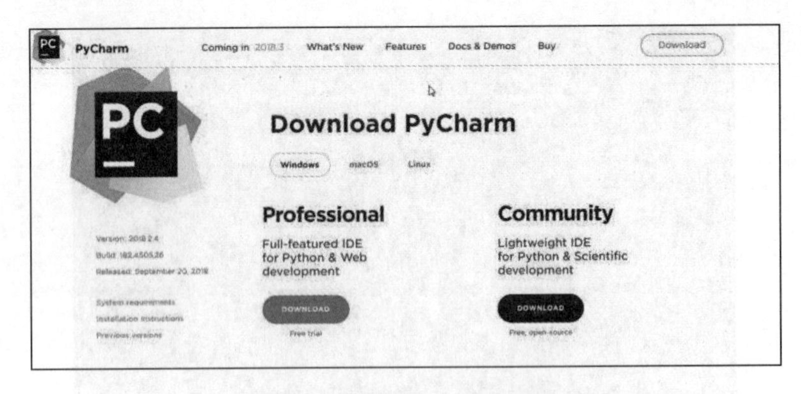

图 1.6 PyCharm 社区版下载

（2）下载安装包之后，双击进行安装，选择安装路径，关联 .py 文件，单击 Next 按钮安装，如图 1.7 所示。

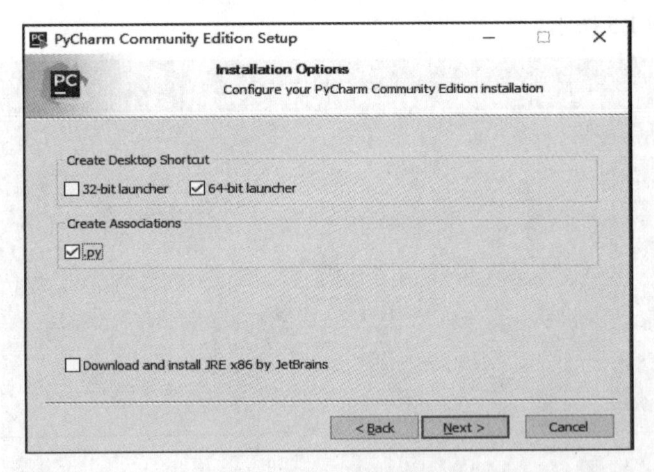

图 1.7 PyCharm 安装选项

（3）一直单击 Next 按钮，即可安装完成，做一些个性化的设置后即可开始创建项目，如图 1.8 所示。

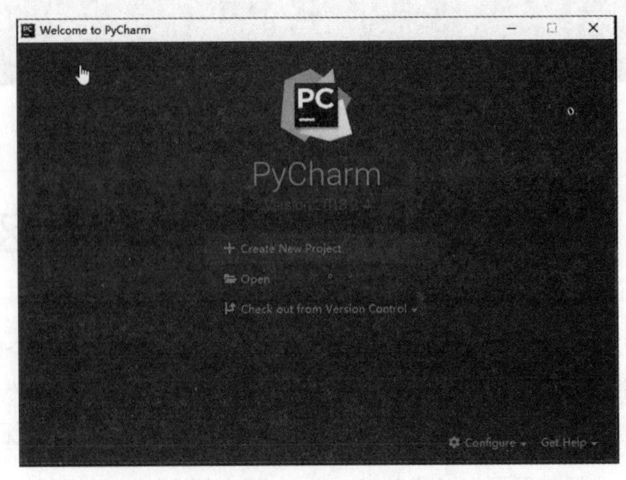

图 1.8 PyCharm 创建项目

1.3 安装开发工具 Visual Studio 社区版

另一个编者使用起来也很方便的编辑器是 Visual Studio 社区版，同样免费，下载地址为 https://visualstudio.microsoft.com/zh-hans/，选择 Community 2017 进行下载。双击下载文件进行安装，由于是通过网络安装，因此需要下载一些安装文件。在安装时选择“Python 开发”复选框，之后开始安装直至完成，如图 1.9 所示。

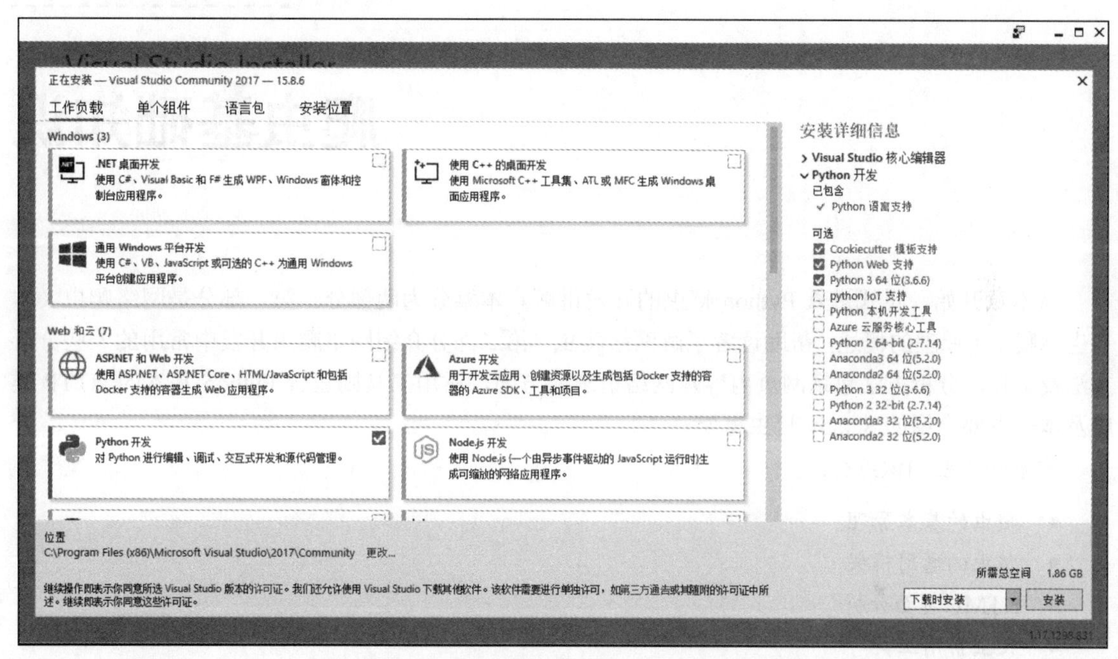

图 1.9　Visual Studio 社区版安装选项

第2章 爬虫基础知识

从本章开始，正式进入Python爬虫的开发讲解。本章分为两部分：第一部分是网络爬虫（本书也称爬虫）原理的概述，帮助读者了解网络爬虫；第二部分介绍网络爬虫开发中常用的一些分析方法及工具，分析方法包括网页内容及网络请求两方面，常用工具则包含Python基本的HTTP类库及本书主要介绍的Scrapy爬虫框架。

本章的主要知识点有：

- 爬虫的基本原理
- 爬虫的通用框架
- HTML页面分析
- 爬虫常用工具

2.1 爬虫原理

网络爬虫在本质上就是模拟用户在浏览器上操作，发送请求，接收响应，然后分析并保存数据，只不过这个过程通过代码实现了大量的自动化操作。

2.1.1 爬虫运行基本流程

一般来说，一个爬虫的执行过程如图2.1所示。

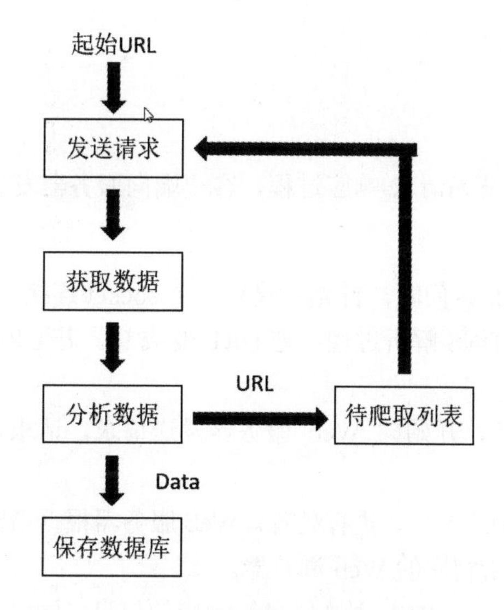

图 2.1　爬虫基本流程

（1）给定 URL，发送 HTTP 请求，即 Request（请求）。
（2）服务器响应请求，得到一个 Response（响应）。
（3）分析返回数据，根据指定规则提取数据，产生 Data 与 URL 两部分数据。
（4）将 Data 保存到数据库、文件等，URL 保存到待爬取列表中，继续爬取。
（5）重复步骤（1）~（4）。

由此可以看出，一个爬虫系统中应包含以下部分。

（1）URL 管理器：负责管理待爬取的网页 URL。
（2）数据下载器：根据 URL 下载数据。
（3）数据分析器：分析筛选下载的数据。
（4）数据保存器：将筛选出的数据保存到文件或数据库。
（5）调度器：负责整个系统的调度。

在发送的 Request 中应包含：

- 请求 URL，需爬取的网页地址。
- 请求方式，爬虫一般用到 POST、GET 两种方式。
- 请求头，一个请求的头部信息，包含 User-Agent、Cookie 等信息。
- 请求体，请求时提交的数据，如登录时的用户名、密码。

相对应的，Response 中则包含所有的响应信息：

- 响应状态，2**代表成功，3**代表重定向，4**客户端错误，5**服务器错误，等等。
- 响应头，包含 Cookie、类型等。
- 响应体，最主要的部分，爬取的数据从中提取，一般类型有网页、图片、文件等。

对数据进行保存时，可以保存到本地文件，如 CSV、Excel、JSON 等，更好的方式是存储到数据库，如 MySQL、MongoDB 等。

2.1.2 HTTP 请求过程

一次 HTTP 请求可以简单理解为请求-响应过程，客户端向服务器发送请求，服务器向客户端返回响应。过程如下。

（1）连接：当我们输入 URL 访问时，首先要建立一个 socket 连接，因为 socket 是通过 IP 和端口建立的，所以之前还有一个 DNS 解析过程，把 URL 变成 IP。若 URL 里不包含端口号，则使用默认端口号。

（2）请求：连接成功建立后，开始向 Web 服务器发送请求，请求常用的是 GET 或 POST 命令。

（3）响应：Web 服务器收到请求后，进行处理。Web 服务器根据请求信息查找文件，如果找到该文件，就把该文件内容传送给相应的 Web 浏览器。

（4）断开连接：当响应结束后，Web 浏览器显示响应的信息，同时与 Web 服务器断开连接。

了解整个请求过程之后，再来看一下请求与响应的内容。

1. 请求内容

一个完整的 HTTP 数据请求包括请求行、请求头、请求数据，如图 2.2 所示。

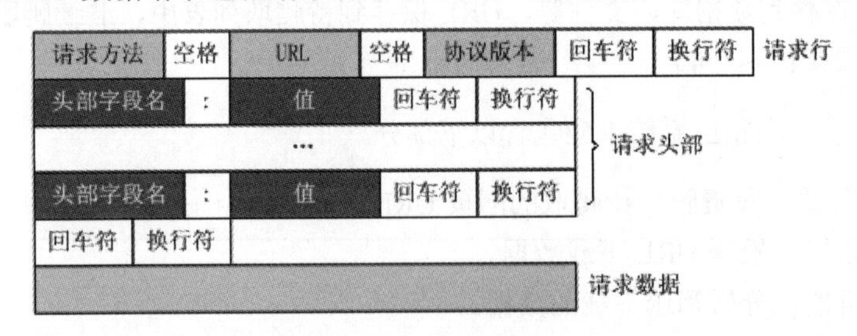

图 2.2　HTTP 请求数据

以如下示例进行说明：

```
01 GET /sample.jsp HTTP/1.1
02 Accept:image/gif.image/jpeg,*/*
03 Accept-Language:zh-cn
04 Connection:Keep-Alive
05 Host:localhost
06 User-Agent:Mozila/4.0(compatible;MSIE5.01;Window NT5.0)
07 Accept-Encoding:gzip,deflate
08
09 username=jinqiao&password=1234
```

上面的代码中，第 01 为请求行，请求方法为 GET，URL 为/sample.jsp，HTTP/1.1 表明使用 HTTP1.1 协议标准。

第 02~07 行为请求头部，做部分说明：Accept 为请求报头域，用户指定客户端接收哪些类型的信息，image/gif 表明接收 GIF 类型的图像信息。Host 头域用于指定请求服务器主机和端口号。

第 09 行是请求正文，请求头和请求正文之间是一个空行（第 08 行），这个行非常重要，它表示请求头已经结束。请求正文中包含客户端提交的查询字符串信息等。

2. 响应内容

响应包含状态行、消息报头与响应正文，代码如下：

```
01 HTTP/1.1 200 OK
02 Date: Mon, 27 Jul 2018 12:28:53 GMT
03 Server: Apache
04 Last-Modified: Wed, 22 Jul 2018 19:15:56 GMT
05 Accept-Ranges: bytes
06 Content-Type: text/plain
07 Cookie: toweibologin=login;
08
09 {''username'': ''Li'', ''password'': ''123456''}
```

上面的代码中，第 01 行是状态行，包含请求协议、响应码、响应状态。

第 02~07 行是响应报文头，同样由多个属性组成，对爬虫而言最常用的就是 Cookie，使用 Cookie 可以直接登录网站，避免网站的登录验证。

第 09 行是服务器返回的内容，形式有 HTML、JSON 等格式。

2.2　网页分析方法 1：浏览器开发人员工具

爬虫作业中，最重要的一步就是提取数据。我们可以根据 HTML 结构来分析页面元素进行提取，也可以根据正则表达式提取对应的数据。分析 HTML 结构就需要用到浏览器开发人员工具。

浏览器开发人员工具可以详细地查看网页布局、请求、响应数据，是爬虫应用中非常实用的工具，也是常用的工具。下面以 Chrome 浏览器中的开发人员工具进行讲解，其他浏览器类似。

打开一个网页，按 F12 快捷键，打开 Developer Tool，如图 2.3 所示。

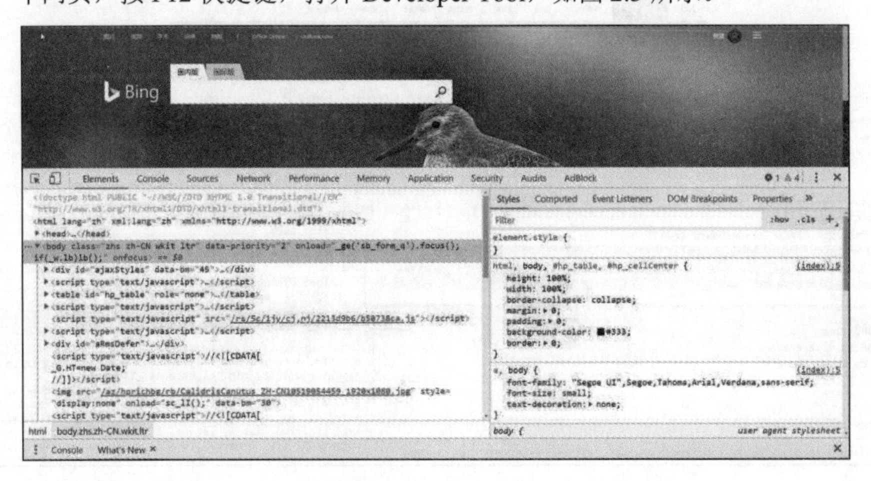

图 2.3　浏览器开发人员工具

可以看到主页面分为上面的工具栏与下面的信息展示栏两部分。工具栏有 8 个主要的工具可以查看开发工具：

- Elements：解析显示页面的组成元素，也就是当前看到 Chrome 渲染页面所需要的 HTML、CSS 和 DOM（Document Object Model）对象。此外，还可以编辑这些内容以更改页面显示结果。
- Console：显示各种警告与错误信息，并且提供了 Shell 用来与文档、开发者工具交互。
- Network：显示当前页面向服务器请求了的资源、资源的大小、加载资源花费的时间以及资源加载是否成功。此外，还可以查看 HTTP 的请求头、请求参数、响应内容等。
- Sources：可以在源代码面板中设置断点来调试 JavaScript，或者通过 Workspaces（工作区）连接本地文件来使用开发者工具进行实时编辑。
- Memory：如果需要比 Performance 提供更多的信息，可以使用 Memory 面板，例如跟踪内存泄漏等。
- Performance：可以通过记录和查看网站生命周期内发生的各种事件来提高页面的运行时性能。
- Application：检查加载的所有资源，包括 IndexedDB 与 Web SQL 数据库、本地和会话存储、Cookie、应用程序缓存、图像、字体和样式表。
- Audits：分析页面加载的过程，进而提供减少页面加载时间、提升响应速度的方案。
- Security：安全面板可以调试混合内容问题、证书问题等。

在爬虫开发中，主要用到的工具有 Elements 与 Network。前者用于查看所需抓取的元素，后者用于查看 HTTP 请求的相关参数与方法。

2.2.1 Elements 面板

选择 Elements 面板，单击工具栏最左侧的鼠标按钮后，单击页面任意元素，在详情区域会被高亮显示，再次单击可退出该模式，如图 2.4 所示。

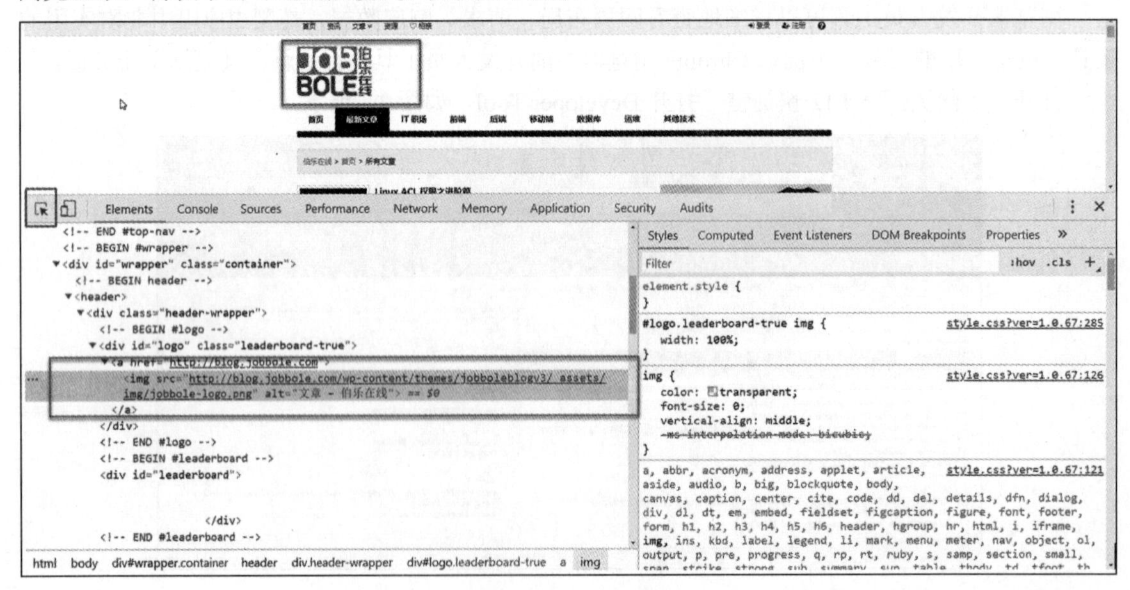

图 2.4 Elements 面板

通过此功能可以快速实现元素的定位，对调试代码非常有用。

在 HTML 页面中，每个元素（如<div>、<a>）都是一个 DOM 节点，所有的 DOM 节点组成了 DOM 树。因此可以把详情区域当作 DOM 树，把 HTML 元素标签看作 DOM 节点。同样，在单击选择 DOM 节点时，网页中对应的元素也会被高亮显示，如图 2.5 所示。

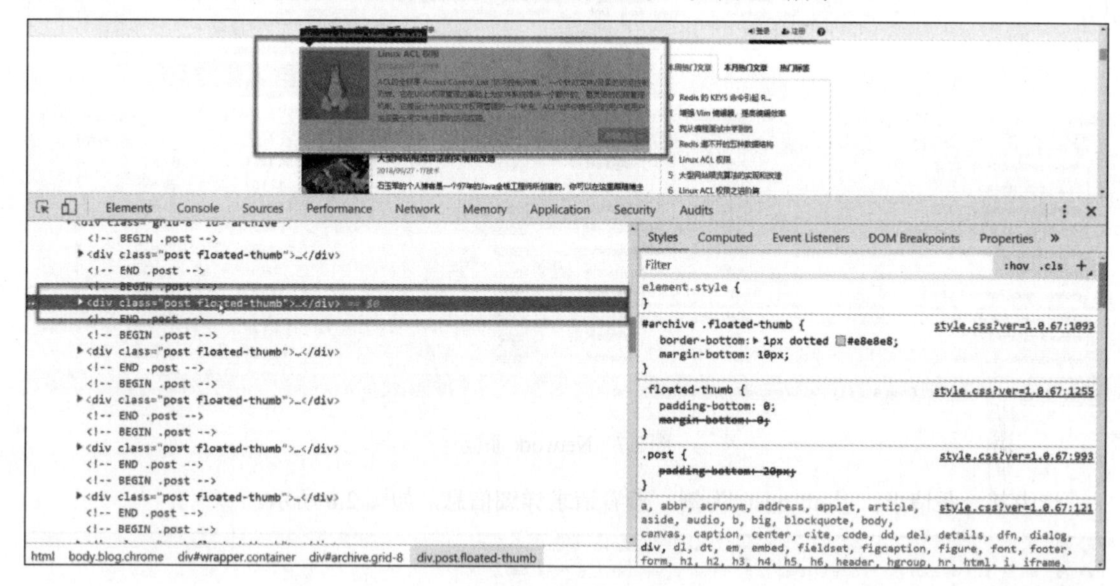

图 2.5　元素定位

找到需要抓取的元素，在代码中一般使用 CSS 与 XPath 两种方式来实现定位。Elements 面板中可以直接右击复制相应的定位方法，Copy selector 为 CSS 路径，Copy XPath 为 XPath 路径，如图 2.6 所示。

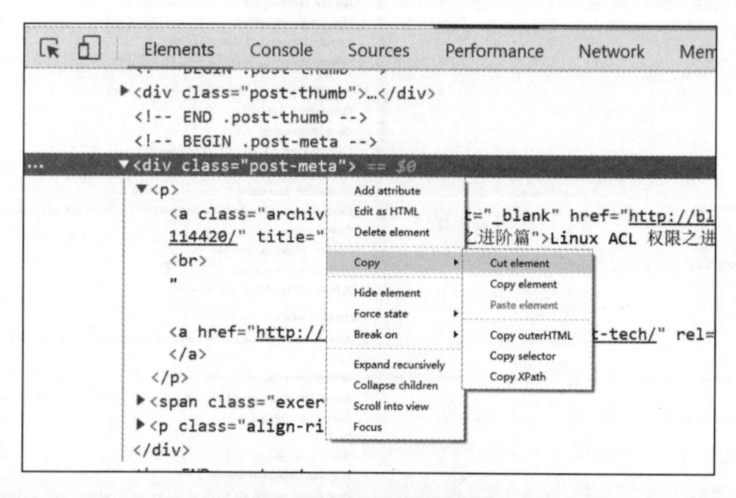

图 2.6　复制元素定位路径

2.2.2　Network 面板

在打开一个网页时，会发起相应的 HTTP 请求，而 Network 面板能够监听记录所有的网络访问请求与响应信息，这一点在分析异步加载请求时非常有用。例如访问伯乐在线左右文章页面，如图 2.7 所示。

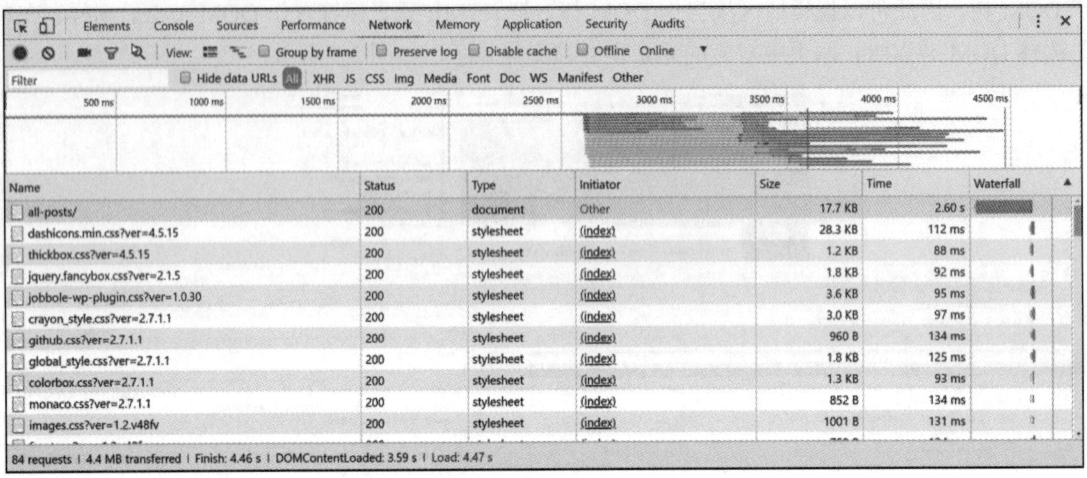

图 2.7　Network 面板

选中某一条请求，以 all-posts/为例，查看请求详细信息，如图 2.8 所示。

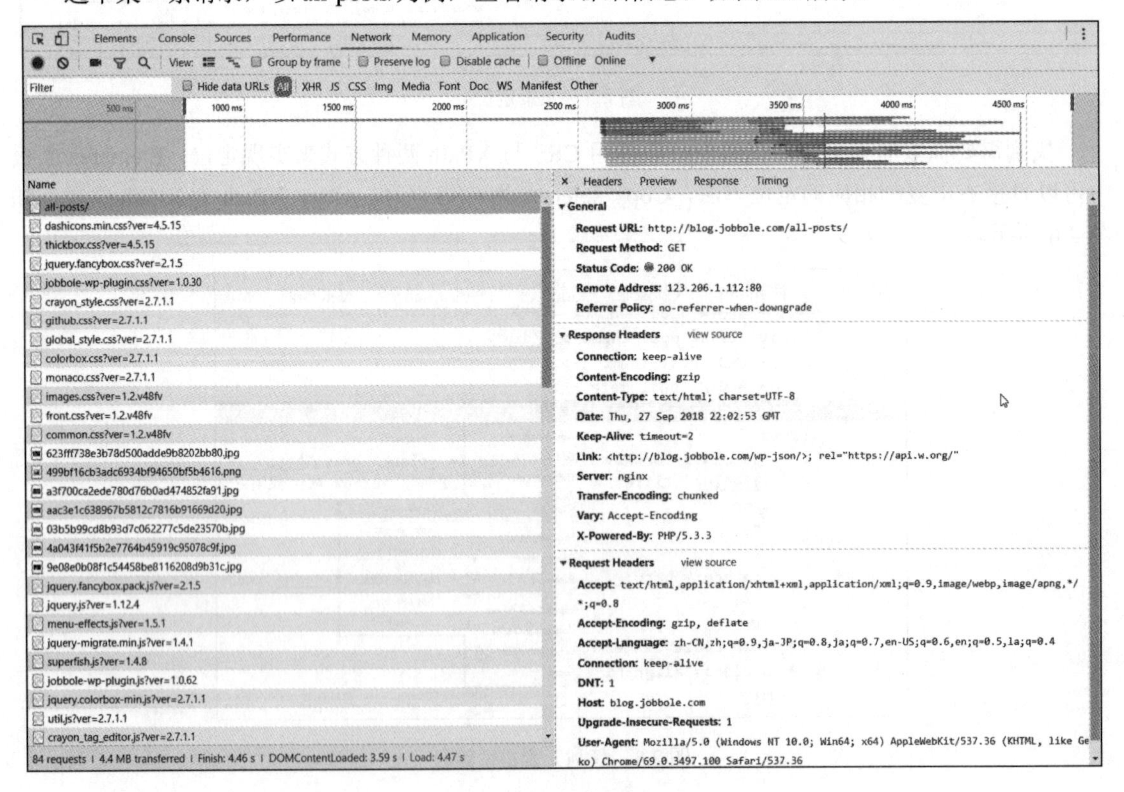

图 2.8　请求数据信息

详细信息中有 Headers、Preview、Response、Timing 四个标签页。

（1）在 Headers 标签中，主要分为以下 3 部分。

- General：HTTP 请求的基本信息，包括请求地址、请求方法、响应状态码、远程 IP、资源访问策略。
- Response Headers：服务器返回的头部信息，包括连接状态、内容返回类型、时间等信息。
- Request Headers：发起请求的头部信息，包括接受类型、接受语言、连接状态等，需要关注的是 User-Agent，某些网站对爬虫类 User-Agent 会有限制，比如 Scrapy 框架默认的 User-Agent，因此需要对该值进行手动修改。

（2）Preview 标签页对返回的信息进行格式化预览，如图 2.9 所示。

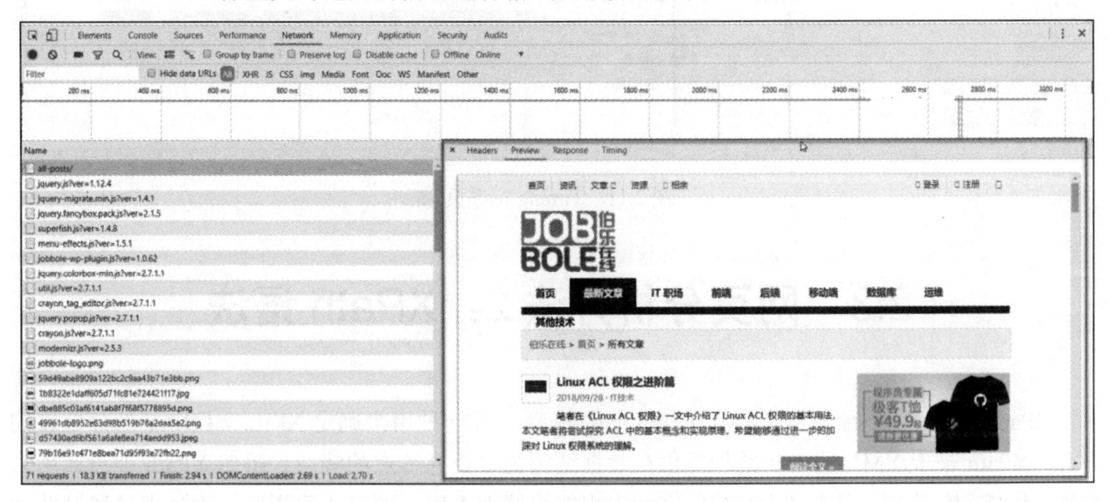

图 2.9　Preview 预览返回信息

（3）Response 标签页则显示相应的信息，如图 2.10 所示。

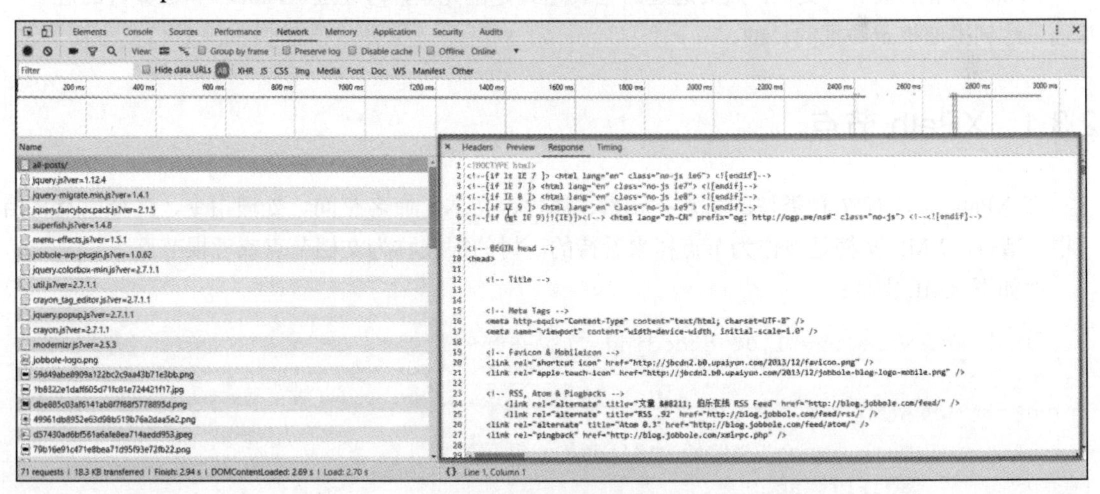

图 2.10　Response 响应数据

（4）Timing 标签页显示每一步动作所耗费的时间，如图 2.11 所示。

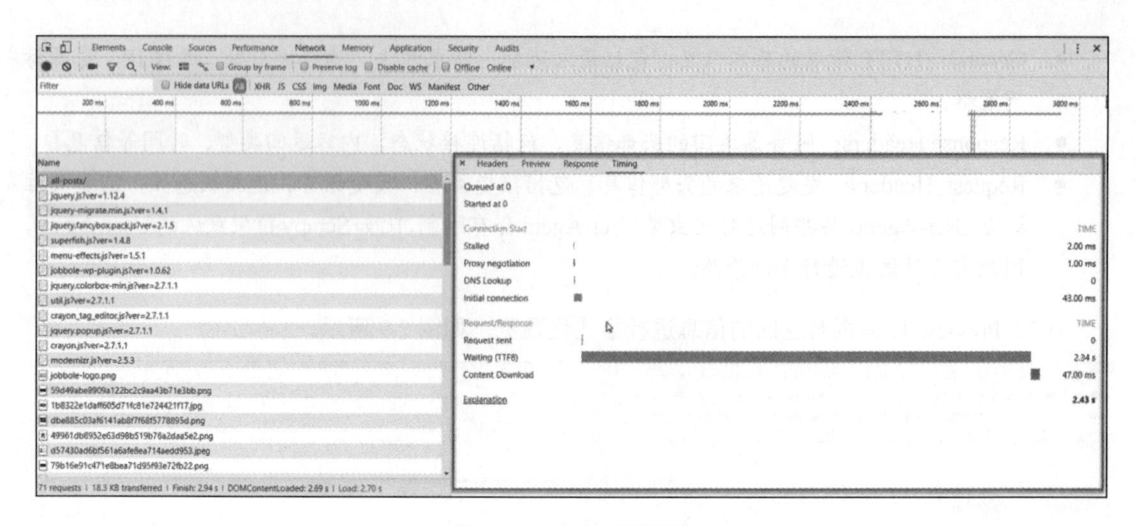

图 2.11 Timing 标签页

2.3 网页分析方法 2：XPath 语法

XPath 为 XML 路径语言（XML Path Language），是一种用来确定 XML 文档中某部分位置的语言。XPath 基于 XML 的树状结构提供在数据结构树中找寻节点的能力。虽然 XPath 被设计用来查找定位 XML 文档，不过在 HTML 文档中也能很好地工作，而且主流浏览器大多支持通过此功能来查找节点元素。因此，在爬虫开发中，必备技能便是通过 XPath 提取网页信息。

XPath 以节点来解析文档，然后通过路径表达式定位元素，再经过 XPath 轴和运算符的进一步筛选，达到提取规定数据的目的。

2.3.1 XPath 节点

在 XPath 中，有 7 种类型的节点：元素、属性、文本、命名空间、处理指令、注释以及文档（根）节点。XML 文档是被作为节点树来看待的。树的根被称为文档节点或者根节点。

有如下 XML 文档：

```
01 <?xml version="1.0" encoding="ISO-8859-1"?>
02 <school>
03 <student>
04       <name lang="en">Make</name>
05       <age>17</age>
06       <gender>M</gender>
07       <score>90</score>
08 </student>
09 </school>
```

上面的 XML 文档中包括的节点如下：

- <school>文档节点
- <name>元素节点
- lang="en"属性节点

节点组合在一起便会产生相对应的节点关系，包括父（Parent）、子（Childen）、同胞（Sibling）、先辈（Ancestor）、后代（Descendant）。

- student 是 name、age、gender 的父，school 是 student 的父。
- name、age、gender 是 student 的子，student 是 school 的子。
- name、age、gender 都是同胞。
- school、student 是 name、age、gender 的先辈，因此，先辈是节点的父、父的父等。
- school 的后代是 name、age、gender、student，也就是节点的子、子的子等。

2.3.2　XPath 语法

XPath 使用路径表达式来选取 XML 文档中的节点或节点集。节点是通过沿着路径（Path）或者步（Steps）来选取的。通过下面的演示文档来看看如何选取节点：

```
01 <?xml version="1.0" encoding="ISO-8859-1"?>
02
03 <school>
04
05   <student>
06     <name lang="eng">Leica</name>
07     <score>97</score>
08   </student>
09
10   <student>
11     <name lang="eng">Make</name>
12     <score>89</score>
13   </student>
14
15 </school>
```

常用的节点选取路径表达式如表 2-1 所示。

表2-1　XPath路径表达式

表达式	描述
nodename	选取此节点的所有子节点
/	从根节点选取
//	从匹配选择的当前节点选择文档中的节点，而不考虑它们的位置
.	选取当前节点
..	选取当前节点的父节点
@	选取属性

通过使用上面的路径表达式，我们对示例文档中的内容进行选取，如表 2-2 所示。

表2-2 节点选取示例

表达式	描述
school	选取 school 元素的所有子节点
/school	选取 school 元素
/school/student	选取 school 的子元素的所有 student 元素
//student	选取所有的 student 元素，而不考虑它们的位置
school//student	选取 school 元素的后代的所有 student 元素，而不考虑它们位于 school 之下的什么位置
@lang	选取名为 lang 的所有属性

表 2-2 中的表达式选取的都是某一类指定的节点，如果要选取指定的某一个节点，就需要用谓语，也就是特殊的限定条件，使用方法如表 2-3 所示。

表2-3 选取指定节点

表达式	描述
/school/student[1]	选取属于 school 子元素的第一个 student 元素
/school/student[last()]	选取属于 school 子元素的最后一个 student 元素
/school/student[last()-1]	选取属于 school 子元素的倒数第二个 student 元素
/school/student[position()<3]	选取最前面的两个属于 school 元素的子元素的 student 元素
//name[@lang]	选取所有拥有名为 lang 的属性的 name 元素
//name[@lang='eng']	选取所有 name 元素，且这些元素拥有值为 eng 的 lang 属性
/school/student[score>90]	选取 school 元素的所有 student 元素，且其中的 score 元素的值需大于 90
/school/student[score>90]/name	选取 school 元素中的 student 元素的所有 name 元素，且其中的 score 元素的值需大于 90

在 XPath 中，使用通配符"*"来匹配未知元素。表 2-4 列出了一些通配元素路径表达式以及匹配结果。

表2-4 "*"通配符

表达式	描述
/school/*	选取 school 元素的所有子元素
//*	选取文档中的所有元素
//name[@*]	选取所有带有属性的 name 元素

如果匹配多个路径，则可以使用"|"运算符，使用方式及匹配结果如表 2-5 所示。

表2-5 "|"运算符

表达式	描述
//student/name | //student/score	选取 student 元素的所有 name 和 score 元素
//name | //score	选取文档中的所有 name 和 score 元素

（续表）

表达式	描述
/school/student/name | //score	选取属于 school 元素的 student 元素的所有 name 元素，以及文档中所有的 score 元素

2.3.3　XPath 轴

轴（Axis）可定义相对于当前节点的节点集。XPath 使用位置路径匹配节点位置，而位置路径类型包括绝对路径与相对路径。绝对路径起始于“/”，相对路径没有限制。在两种情况中，位置路径均包括一个或多个步，每个步均被斜杠分割：绝对路径/step/step/...，相对路径 step/step/...。每个步均根据当前节点集中的节点进行计算，轴的存在可使节点提取变得更加灵活准确。

XPath 轴中的节点集如表 2-6 所示。

表2-6　XPath轴节点

节点	描述
ancestor	选取当前节点的所有先辈（父、祖父等）
ancestor-or-self	选取当前节点的所有先辈（父、祖父等）以及当前节点本身
attribute	选取当前节点的所有属性
child	选取当前节点的所有子元素
descendant	选取当前节点的所有后代元素（子、孙等）
descendant-or-self	选取当前节点的所有后代元素（子、孙等）以及当前节点本身
following	选取文档中当前节点的结束标签之后的所有节点
namespace	选取当前节点的所有命名空间节点
parent	选取当前节点的父节点
preceding	选取文档中当前节点的开始标签之前的所有节点
preceding-sibling	选取当前节点之前的所有同级节点
self	选取当前节点

步包括：

- 轴（Axis），定义所选节点与当前节点之间的树关系。
- 节点测试（Node-Test），识别某个轴内部的节点。
- 零个或者更多谓语（Predicate），更深入地提炼所选的节点集。

步的语法为：轴名称::节点测试[谓语]。

使用的测试文档如下，通过此文档进行轴的示例分析：

```
01 <?xml version="1.0" encoding="ISO-8859-1"?>
02
03 <school>
04
05   <studentA>
```

```
06     <name lang="eng">Leica</name>
07     <score>97</score>
08   </studentA>
09
10   <studentA>
11     <name lang="eng">Make</name>
12     <score>89</score>
13   </studentA>
14
15   <studentB>
16     <name lang="eng">Make</name>
17     <score>88</score>
18   </studentB>
19
20   <studentB>
21     <name lang="eng">Make</name>
22     <score>90</score>
23   </studentB>
24
25 </school>
```

通过演示文档操作示例如表 2-7 所示。

表2-7　步的使用示例

示例	说明
//store/ancestor::*	选取当前 score 节点的所有先辈（父、祖父等）
//name/ancestor::student	选择当前 name 节点的所有 student 先辈
//score/ancestor-or-self::*	选取当前 score 节点的所有先辈（父、祖父等）以及当前节点本身
/school/studentA/name/attribute::*	选取当前 name 节点的所有属性
//name/attribute::lang	选取当前 name 节点的 lang 属性
//studentA/child::*	选取当前 studentA 节点的所有子元素
//score/child::text()	选取当前 score 节点的所有文本子节点
//studentA/child::node()	选取当前 studentA 节点的所有子节点
school/child::*/child::score	选取当前 school 节点的所有 score 孙节点
/school/descendant::*	选取当前 school 节点的所有后代元素（子、孙等）
/school/descendant-or-self::*	选取当前 school 节点的所有后代元素（子、孙等）以及当前节点本身
/school/studentA[2]/following::*	选取文档中当前第二个 studentA 节点之后的所有节点，也就是所有 studentB 节点及其子节点
/school/studentB[1]/preceding::*	选取文档中当前第一个 studentB 节点之前的所有节点，也就是所有 studentA 节点及其子节点
/school/studentA[2]/following-sibling::*	选取当前第二个 studentA 节点之前的所有同级节点，也就是所有的 studentB 节点

（续表）

示例	说明
/school/studentB[1]/preceding-sibling::*	选取当前第一个 studentB 节点之前的所有同级节点，也就是所有的 studentA 节点
//studentA/self::*	选取当前 studentA 节点

2.3.4　XPath 运算符

XPath 表达式可返回节点集、字符串、逻辑值以及数字。表 2-8 列出了可用在 XPath 表达式中使用的运算符。

表2-8　XPath运算符

运算符	说明	示例	示例说明
|	计算两个节点集	//studentA | //studentB	返回所有拥有 studentA 和 studentB 元素的节点集
+	加法	//studentA[score=80+10]	选取子元素 score 等于 90 的 studentA 元素
-	减法	//studentA[score=100-10]	选取子元素 score 等于 90 的 studentA 元素
*	乘法	//studentA[score=9*10]	选取子元素 score 等于 90 的 studentA 元素
div	除法	//studentA[score=180 div2]	选取子元素 score 等于 90 的 studentA 元素
=	等于	//studentA[score=90]	选取子元素 score 等于 90 的 studentA 元素
!=	不等于	//studentA[score!=80]	选取子元素 score 不等于 80 的 studentA 元素
<	小于	//studentA[score<100]	选取子元素 score 小于 100 的 studentA 元素
<=	小于或等于	//studentA[score<=80]	选取子元素 score 小于等于 80 的 studentA 元素
>	大于	//studentA[score>80]	选取子元素 score 大于 80 的 studentA 元素
>=	大于或等于	//studentA[score>=80]	选取子元素 score 大于等于 80 的 studentA 元素
or	或	//studentA[score<80 or score>90]	选取子元素 score 小于 80 或者 score 大于 90 的 studentA 元素
and	与	//studentA[score>80 and score<90]	选取子元素 score 大于 80 并且小于 90 的 studentA 元素
mod	计算除法的余数	//studentA[score=190 mod 100]	选取子元素 score 等于 90 的 studentA 元素

2.4　网页分析方法 3：CSS 选择语法

CSS 指层叠样式表（Cascading Style Sheets），用来定义如何显示 HTML 元素，一般和 HTML 文件一起使用，通过对 HTML 标签的修饰来解决内容与表现分离的问题。

CSS 规则由两个主要的部分构成：选择器以及一条或多条声明，具体如下：

```
selector {declaration1; declaration2; ... declarationN }
```

选择器通常是需要改变样式的 HTML 元素，需要指定具体的标签。每条声明由一个属性和一个值组成。属性（Property）是我们希望设置的样式属性（Style Attribute）。每个属性有一个值。属性和值被冒号分开。在进行爬虫开发时，主要通过选择器来进行相关元素的提取，因此我们主要对选择器的相关用法做讲解，其他的内容如有兴趣，读者可自行查看相关资料。

CSS 的选择器主要包括元素选择器、类选择器、ID 选择器、属性选择器、后代选择器、子元素选择器、相邻兄弟选择器。下面将根据如下文档进行具体讲解：

```
01 <!DOCTYPE html>
02 <html lang="en" dir="ltr">
03   <head>
04     <meta charset="utf-8">
05     <title>考试说明</title>
06   </head>
07   <body>
08     <h1>考试说明</h1>
09     <p>2018 年第二学期高数考试说明</p>
10     <div class="info">
11       <p>考试因故取消</p>
12         <a id="detail" href="/detail/info.html">
13           <img src="/cancel.jpg" alt="">
14         查看详情
15         </a>
16     </div>
17     <p class="info">下次补考时间:2019.1.3</p>
18     <p target="freshman">一年级</p>
19   </body>
20 </html>
```

2.4.1 元素选择器

最常见的 CSS 选择器是元素选择器。也就是说，文档中的元素就是基本的选择器，比如 p、h1、div、a，如表 2-9 所示。

表2-9 CSS元素选择器

选择器	示例	描述
element	p	选择所有<p>元素
element	div	选择所有<div>元素

2.4.2　类选择器

类选择器允许以一种独立于文档元素的方式来指定元素，该选择器可以单独使用，也可以与其他元素结合使用，如表 2-10 所示。

表2-10　CSS类选择器

选择器	示例	描述
.class	.info	选择所有 class="info"的元素
element.class	div.info	选择所有 class="info"的 div 元素

2.4.3　ID 选择器

ID 选择器与类选择器类似，不同的是 ID 选择器以“#”开始，并且由于在 HTML 文档中 ID 值是唯一的，因此比 class 更方便使用，但并不是每个标签都有 ID，使用方法如表 2-11 所示。

表2-11　CSS ID选择器

选择器	示例	描述
#id	#detail	选择所有 id="detail"的元素
element#id	a#detail	选择所有 id="detail"的 a 元素

2.4.4　属性选择器

如果希望选择有某个属性的元素，则可以使用属性选择器，不仅限于 class 和 id 属性，如表 2-12 所示。

表2-12　CSS属性选择器

选择器	示例	描述
[attribute]	[target]	选择所有含有 target 属性的元素
[attribute=value]	[target="freshman"]	选择 target 属性等于"freshman"的元素
[attribute~=value]	[target~="freshman"]	选择 target 属性包含 freshman 的元素
[attribute^=value]	[target^="fresh"]	选择 target 属性以 fresh 开头的元素
element[attribute]	a[attribute]	选择所有含有 target 属性的 a 元素

2.4.5　后代选择器

后代选择器（Descendant Selector）又称为包含选择器，可以选择作为某元素后代的元素，如表 2-13 所示。

表2-13 CSS后代选择器

选择器	示例	描述
element element	div p	选择 div 元素下的所有 p 元素
element.class element	div.info img	选择 class 等于 info 的 div 元素下的所有 img 元素

2.4.6 子元素选择器

如果不希望选择任意的后代元素，而是希望缩小范围，只选择某个元素的子元素，可以使用子元素选择器（Child Selector），如表 2-14 所示。

表2-14 CSS子元素选择器

选择器	示例	描述
element > element	div > p	选择 div 元素下的所有 p 子元素

2.4.7 相邻兄弟选择器

如果需要选择紧接在另一个元素后的元素，而且二者有相同的父元素，可以使用相邻兄弟选择器（Adjacent Sibling Selector），如表 2-15 所示。

表2-15 CSS相邻兄弟选择器

选择器	示例	描述
element + element	p + a	选择紧跟在 p 元素后面的所有兄弟 a 元素

单独使用某一种选择器实现选取指定元素的情况并不常见，多数情况下需要组合使用，如表 2-16 所示。

表2-16 CSS选择器组合使用

示例	描述
html > body div.info + p	选择紧接在 class 等于 info 的 div 元素后出现的所有兄弟 p 元素，该 div 元素包含在一个 body 元素中，body 元素本身是 html 元素的子元素

2.5 网页分析方法 4：正则表达式

对于复杂的数据结构，或者无法使用 XPath、CSS 进行提取的数据，使用正则表达式是一个很好的选择。正则表达式就是使用一些特殊规定的字符组成包含一定逻辑的特殊组合，使用这个特殊组合对字符串进行过滤，找到满足该逻辑的数据。

2.5.1　提取指定字符

获取数据结合，有用的数据只有一部分，我们可以指定字符串进行匹配，如表 2-17 所示。

表2-17　字符匹配

语法	含义	表达式	匹配的字符串
一般字符串	匹配自身	abc	abc
.	可以匹配除了换行符“\n”之外的任意字符	a.c	abc、aec、a#c、a9c
[]	匹配中括号内的任意字符	a[1b$]c	a1c、abc、a$c
[^]	“^”在中括号内，表示取反，即匹配除了中括号内的数据	a[^123]c	abc、a4c
\	转义字符，使后一个字符改变原来的意思，如“\.”中的“.”代表本身，而不是任意字符	a[db\.\\]c	adc、abc、a.c、a\c

2.5.2　预定义字符集

数字信息应该是爬虫工作中最频繁提取的数据了，如日期、交易量、订单数等。用来匹配数字的字符是\d，而提取文字信息一般用\w，这种叫作预定义字符集，常用的预定义字符集如表 2-18 所示。

表2-18　预定义字符集

语法	含义	表达式	匹配的字符串
^	匹配字符串的开始	^abc	abcde
$	匹配字符串的结尾	abc$	123abc
\b	匹配单词的开始或结束	\babc\b	abc
\d	匹配数字 0~9	a\dc	a3c
\D	匹配非数字，相当于[^\d]	a\Dc	a@c
\s	匹配任意空白字符，等价于 [<空格>\t\n\r\f].	a\sc	a c
\S	匹配任意非空白字符，等价于[^\s]	a\Sc	aDc、a*c
\w	匹配单词字符[A-Za-z0-9_]，字母、数字、下画线	a\wc	abc、a1c、aBc、a_c
\W	匹配非单词字符，等价于[^\w]	a\Wc	a%c、a c、a-c

2.5.3　数量限定

当需要重复匹配时，比如电话号码 11 个数字，匹配 11 次\d，为了避免写 11 次\d 或者更多次，就需要用到限定符，如表 2-19 所示。

表2-19 数量限定

语法	含义	表达式	匹配的字符串
*	匹配前一个字符 0 次或多次	abc*	ab、abccc
+	匹配前一个字符 1 次或多次	abc+	abc、abccc
?	匹配前一个字符 0 次或 1 次	abc?	ab、abc
{m}	匹配前一个字符 m 次	a{2}bc	aabc
{m,}	匹配前一个字符 m 次或更多次	a{2,}bc	aabc、aaaabc
{,m}	匹配前一个字符 0-m 次	a{,2}bc	bc、abc、aabc
{m,n}	匹配前一个字符 m 至 n 次	a{1,3}bc	abc、aabc、aaabc

2.5.4 分支匹配

如果有多种匹配规则，则满足任意一种规则即可完成匹配，这时就要用到分支条件的匹配。比如日期格式，2018-10-10、2018/10/10 或者 2018.10.10 都满足条件，从左往右，匹配到任意一个即可，如表 2-20 所示。

表2-20 条件分支

<table>
<tr><th>语法</th><th>含义</th><th>表达式</th><th>匹配的字符串</th></tr>
<tr><td>|</td><td>左右两个表达式满足一个即可，先左后右，一旦左边匹配成功，则跳过右边匹配</td><td>\d{4}-\d{2}-\d{2}|\d{4}/\d{2}/\d{2}|\d{4}\.\d{2}\.\d{2}</td><td>2018-12-12</td></tr>
</table>

2.5.5 分组

表 2-20 中的正则表达式不仅能匹配到 2018-12-12 这样的日期，也能匹配到 2018-99-99 这样不存在的日期，需要对月份日期做具体的限定，月份应该是 0[1-9]|1[0-2]，日应该是 0[1-9]|[12][0-9]|3[01]，完整表达式为\d{4}-(0[1-9]|1[0-2])-(0[1-9]|[12][0-9]|3[01])。这里没有判断每月是否有 31 日，二月是否为闰月，更详细的判断请读者练习完成。

每个“()”为一个分组，分组的起始编号为 1，每遇到一个“(”编号加 1，在提取数据时，可以方便地将一组数据单独提取出来。

2.5.6 零宽断言

先介绍什么是零宽断言，在程序设计中，断言指的是判断一个情况是否为真，而在正则表达式中，零宽断言指的是当断言表达式为真的时候，再进行匹配，匹配的时候并不匹配断言表达式的内容。断言有不同的形式，结合如下字符串更好理解：

```
Rita repeated what Reardon recited when
```

（1）先行断言，也叫作零宽度正预测先行断言，表达式：(?=exp)，当断言表达式为真时，匹配断言表达式前面的位置。先行断言的执行步骤为：先从要匹配的字符串中的最右端找到第一个断言表达式，再匹配其前面的表达式；若无法匹配，则继续查找第二个断言表达式，再匹配第二个断言表达式前面的字符串；若能匹配，则匹配。例如\w+(?=ted)，会匹配到 repea 和 reci（断言表达是"ted"不会匹配），而.*(?=t)则会匹配到 My WeChat account。

（2）后发断言，也叫作零宽度正回顾后发断言，表达式：(?<=exp)，当断言表达式为真时，匹配断言表达式后面的位置。与先行断言执行顺序相反，先从要匹配的字符串中的最左端找到第 1 个断言表达式，再匹配其后面的表达式；若无法匹配，则继续查找第 2 个断言表达式，再匹配第 2 个断言表达式后面的字符串；若能匹配，则匹配。例如(?<=re)\w+，会匹配到 peated 和 cited，而(?<=re).*则会匹配到 peated what Reardon recited when。

（3）零宽度负预测先行断言，表达式(?!exp)，当断言表达式不成立时，匹配断言表达式前面的位置；若断言表达式成立，则不匹配。例如\b(?!R)\w+匹配不以大写字母 R 开头的字符串。

（4）零宽度负回顾后发断言，表达式(?<!exp)，当断言表达式不成立时，匹配断言表达式后面的位置；若断言表达式成立，则不匹配。例如\w+(?<!ted)\b 匹配不以 ted 结尾的字符串。

2.5.7　贪婪模式与非贪婪模式

正则表达式默认是贪婪模式，也就是尽可能多地匹配字符串，比如使用 a\w+b 匹配 a123b123b 时，得到的是 a123b123b，而不是 a123b。想匹配到 a123b 的时候，就需要启用非贪婪模式，将表达式改为 a\w+?b 就可以匹配到 a123b。数量限定中启用非贪婪模式如表 2-21 所示。

表2-21　数量限定与非贪婪模式

语法	含义
*?	匹配前一个字符 0 次或多次，尽可能少重复
+?	匹配前一个字符 1 次或多次，尽可能少重复
??	匹配前一个字符 0 次或 1 次，尽可能少重复
{m,}?	匹配前一个字符 m 次或更多次，尽可能少重复
{m,n}?	匹配前一个字符 m 至 n 次，尽可能少重复

2.5.8　Python 中的正则表达式

在其他编程语言中，处理正则表达式时，反斜杠"\"时常带来想不到的造成困扰，反斜杠既可作为普通字符，又兼任着转义字符的功能，假如你需要匹配文本中的字符"\"，那么使用编程语言表示的正则表达式需要使用 4 个反斜杠"\\\\"：前两个和后两个分别用于在编程语言中转义成反斜杠，转换成两个反斜杠后，再在正则表达式中转义成一个反斜杠。而在 Python 中，原生字符串很好地解决了这个问题，简单地使用 r"\\"即可。同样，匹配一个数字的"\\d"可以写成 r"\d"。使用原生字符串，无须担心繁杂的转义问题，写出来的表达式也更加直观。

Python 中使用 re 模块来提供对正则表达式的支持。使用 re 的一般步骤是先将正则表达式的字符串形式编译为 Pattern 实例，然后使用 Pattern 实例处理文本并获得匹配结果（一个 Match 实例），最后使用 Match 实例获得信息，进行其他的操作。下面对主要用到的方法进行示例讲解。

（1）re.compile(pattern,flags=0)

将字符串 pattern 转化为 Pattern 对象，用来给其他的 re 函数提供正则表达式参数，做进一步的搜索，其中的 flags 是匹配模式，可选的有：

- re.I：忽略大小写。
- re.M：多行模式，改变'^'和'$'的行为。
- re.S：点任意匹配模式，改变“.”的行为。
- re.L：使预定字符类 \w \W \b \B \s \S 取决于当前区域设定。
- re.U：使预定字符类 \w \W \b \B \s \S \d \D 取决于 Unicode 定义的字符属性。
- re.V：详细模式，这个模式下正则表达式可以是多行的，忽略空白字符，并可以加入注释。

【示例 2-1】使用 re.compile()转化 Pattern 对象：

```
01 import re
02
03 # 将正则表达式编译成 Pattern 对象
04 pattern = re.compile(r'\w+')
```

（2）re.match(pattern, string, flags=0)

从字符串 string 起始位置开始匹配 pattern，若匹配到，则返回一个 Match 对象；若未匹配到，则返回 None。Match 对象有很多方法，最常用的是 group，示例如下。

【示例 2-2】re.match()查找字符串

```
01 import re
02
03 # 将正则表达式转化成 Pattern 对象
04 pattern = re.compile(r'(\d{4})-(\d{8})')
05 # 待匹配字符串
06 string1 = "0755-44445555 is our new office phone number"
07 string2 = "the old number 0755-11112222 is no longer used"
08 #生成 Match 对象
09 match1 = re.match(pattern,string1)
10 match2 = re.match(pattern,string2)
11
12 if match1:
13     # re.match 结果类型
14     print(type(match1))
15     # groups 返回所有匹配结果
16     print(match1.groups())
17     # group(0)为未分组的原始匹配对象
18     print(match1.group(0))
19     # group(1)为第一组对象，以此类推
```

```
20      print(match1.group(1))
21      print(match1.group(2))
22  else:
23      print('match1 未匹配到对象')
24
25  if match2:
26      print(match2)
27      # groups 返回所有匹配结果
28      print(match2.groups())
29      # group(0)为未分组的原始匹配对象
30      print(match2.group(0))
31      # group(1)为第一组对象，以此类推
32      print(match2.group(1))
33      print(match2.group(2))
34  else:
35      print('match2 未匹配到对象')
```

运行结果如下：

```
re.match 返回结果类型为: <class '_sre.SRE_Match'>
groups 的内容为: ('0755', '44445555')
group(0)的内容为: 0755-44445555
group(1)的内容为: 0755
group(2)的内容为: 44445555
match2 未匹配到对象
```

（3）re.search(pattern, string, flags=0)

search 方法与 match 方法类似，不同的是，match 必须在字符串起始位置开始匹配，而 search 则会查找整个 string 进行匹配。若在任意位置匹配到，则返回一个 Match 对象，示例如下。

【示例 2-3】re.search()全文查找字符串

```
01  import re
02
03  # 将正则表达式转化成 Pattern 对象
04  pattern = re.compile(r'(\d{4})-(\d{8})')
05  # 待匹配字符串
06  string1 = "0755-44445555 is our new office phone number"
07  string2 = "the old number 0755-11112222 is no longer used"
08  #生成 search 对象
09  search1 = re.search(pattern,string1)
10  search2 = re.search(pattern,string2)
11
12  if search1:
13      # re.search 结果类型
14      print('search1 返回结果类型为:',type(search1))
15      # groups 返回所有匹配结果
```

```
16      print('search1 中 groups 的内容为:',search1.groups())
17      # group(0)为未分组的原始匹配对象
18      print('search1 中 group(0)的内容为:',search1.group(0))
19      # group(1)为第一组对象，以此类推
20      print('search1 中 group(1)的内容为:',search1.group(1))
21      print('search1 中 group(2)的内容为:',search1.group(2))
22  else:
23      print('search1 未匹配到对象')
24
25  if search2:
26      print('search2 返回结果类型为:',type(search1))
27      # groups 返回所有匹配结果
28      print('re.search 返回结果类型为:',search2.groups())
29      # group(0)为未分组的原始匹配对象
30      print('search2 中 groups 的内容为:',search2.group(0))
31      # group(1)为第一组对象，以此类推
32      print('search2 中 group(0)的内容为:',search2.group(0))
33      print('search2 中 group(1)的内容为:',search2.group(1))
34      print('search2 中 group(2)的内容为:',search2.group(2))
35  else:
36      print('search2 未匹配到对象')
```

运行结果如下：

```
search1 返回结果类型为: <class '_sre.SRE_Match'>
search1 中 groups 的内容为: ('0755', '44445555')
search1 中 group(0)的内容为: 0755-44445555
search1 中 group(1)的内容为: 0755
search1 中 group(2)的内容为: 44445555
search2 返回结果类型为: <class '_sre.SRE_Match'>
re.search 返回结果类型为: ('0755', '11112222')
search2 中 groups 的内容为: 0755-11112222
search2 中 group(0)的内容为: 0755-11112222
search2 中 group(1)的内容为: 0755
search2 中 group(2)的内容为: 11112222
```

（4）re.findall(pattern, string, flags=0)

搜索整个字符串，以列表形式返回所有匹配结果，示例如下。

【示例 2-4】re.findall()查找所有匹配对象

```
01  import re
02
03  # 将正则表达式转化成 Pattern 对象
04  pattern = re.compile('\d+')
05  # 待匹配字符串
06  strings = 'Your activation code is 73829-72993-00983-84721'
```

```
07 result = re.findall(pattern,strings)
08
09 print(result)
```

运行结果如下：

```
['73829', '72993', '00983', '84721']
```

（5）re.split(pattern, string, maxsplit=0, flags=0)

以匹配到的字符串分割 string，返回分割后的列表，示例如下。

【示例 2-5】re.split()分割查找字符串

```
01 import re
02
03 # 将正则表达式转化成 Pattern 对象
04 pattern = re.compile('\W')
05 # 待匹配字符串
06 strings = 'This$is&the@largest%ball'
07
08 result = re.split(pattern,strings)
09
10 print(result)
```

运行结果如下：

```
['This', 'is', 'the', 'largest', 'ball']
```

（6）re.sub(pattern, repl, string, count=0, flags=0)

使用 repl 替换匹配到的字符串，并返回替换后的字符串。repl 可以是一个字符串或者一个方法，当是一个方法时，接收一个 Match 参数并返回一个字符串，使用此字符串进行替换操作，count 指定次数，默认为 0 时全部替换，示例如下。

【示例 2-6】re.sub()替换匹配对象

```
01 import re
02
03 # 将正则表达式转化成 Pattern 对象
04 pattern = re.compile(r'(\d{4}-\d{2}-\d{2})')
05 strings = 'Today is 2018-12-12, the date of the meeting is set at 2019-09-10,
please confirm 06          whether to participate before 2018-12-25'
07
08 # 定义日期格式转换方法，将'-'用'.'代替
09 def totype(match):
10     return match.group(0).replace(r'-','.')
11
12 new_strings = re.sub(pattern,totype,strings)
13
14 print(new_strings)
```

运行结果如下：

```
Today is 2018.12.12, the date of the meeting is set at 2019.09.10, please confirm whether to participate before 2018.12.25
```

2.6 爬虫常用类库 1：Python 中的 HTTP 基本库 urllib

我们已经了解到，爬虫简单来说就是发送请求并获取信息。道理很简单，但怎么发送请求呢？幸运的是，使用 Python 标准库和第三方库都能实现此功能。接下来我们将介绍这些标准库及第三方库。urllib 是 Python 内置的标准库，无须安装即可使用。

2.6.1 发送请求

基本的网络请求方法为 urlopen，方法原型为：

```
urllib.request.urlopen(url, data=None, [timeout, ]*, cafile=None, capath=None, cadefault=False, context=None)
```

参数说明如下。

- url：访问的地址，也可以是一个 Request。
- data：若指定 data 值，则变为 POST 请求（注意：data 传递的参数需要转为 bytes，可通过 urllib.parse.urlencode 进行转换）。
- timeout：设置网站的请求超时时间。
- cafile：在访问 HTTPS 网站时，可使用此参数指定单个 CA 证书文件。
- capath：在访问 HTTPS 网站时，可使用此参数指定 CA 证书文件路径。
- cadefault：此参数已废弃。
- context：必须指定为描述 SSL 选项的 ssl.SSLContext 实例。

urlopen()返回对象可用的方法如下。

- read()：返回响应信息。
- geturl()：返回响应的网址。
- getcode()：返回响应码。
- info()：返回页面元信息，比如头部信息等。

打开一个无参数请求，默认为 GET 请求：

```
>>> import urllib.request
>>> url = 'http://bing.com'
>>> r=urllib.request.urlopen(url)
```

如果发送的 GET 请求有参数，就需要用到 urllib.parse 中的 urlencode 将请求参数进行 URL 编码，拼接之后再进行发送：

```
>>> import urllib.request
>>> import urllib.parse
>>> url = 'http://bing.com/search'
>>> data={'q':'python'}
>>> req_data = urllib.parse.urlencode(data)
>>> req_data
'q=python'
>>> req_url = url +'?' + req_data
>>> req_url
'http://bing.com/search?q=python'
>>> r=urllib.request.urlopen(url)
>>> r.getcode()
200
>>> r.geturl()
'http://cn.bing.com/?scope=web&setmkt=zh-CN&setmkt=zh-CN&setmkt=zh-CN'
>>> r.info()
<http.client.HTTPMessage object at 0x0000018EB53E6E10>
```

如果要发送 POST 请求，就需要加上 data 参数：

```
data = bytes(urllib.parse.urlencode({name: Li}), encoding= 'utf8')
r = urllib.request.urlopen('http://example.com', data=data)
```

2.6.2 使用 Cookie

Cookie 是当浏览某网站时，网站存储的信息文件，它记录了用户的 ID、密码、浏览记录、浏览时间等信息。当同一用户再次访问该网站时，网站通过读取 Cookie，获取用户的相关信息，比如用户名、密码，用于自动登录。Cookie 在爬虫操作中很常用，特别是针对一些需要登录才能抓取信息的网站的抓取工作。使用 urllib 操作网站 Cookie 需要用到 urllib.request.HTTPCookieProcessor(cookie)。使用 Cookie 需要创建一个 opener。在 Python 的 http 包中包含 cookiejar 模块，用于提供对 Cookie 的支持。http.cookiejar 功能强大，我们可以利用本模块的 CookieJar 类的对象来捕获 Cookie，并在后续连接请求时重新发送，比如可以实现模拟登录功能。使用方法如下：

【示例 2-7】 urllib 获取 Cookie

```
01 # 获取 Cookie
02 import http.cookiejar, urllib.request
03 cookie = http.cookiejar.CookieJar()
04 handler = urllib.request.HTTPCookieProcessor(cookie)
05 opener = urllib.request.build_opener(handler)
06 response = opener.open('http://www.baidu.com')
07 for item in cookie:
08     print(item.name+"="+item.value)
```

输出结果如图 2.12 所示。

```
BAIDUID=A5E506AA5397AA1CC5B7DF6233B7CBFD:FG=1
BIDUPSID=A5E506AA5397AA1CC5B7DF6233B7CBFD
H_PS_PSSID=26525_1443_21078_28205_28131_26350_27245_22158
PSTM=1546609437
delPer=0
BDSVRTM=0
BD_HOME=0
```

图 2.12 获取 Cookie

【示例 2-8】urllib 保存 Cookie

```
01 # 保存 Cookie
02 filename = 'saved_cookies.txt'
03 # FileCookieJar、MozillaCookieJar、LWPCookieJar 均为保存 Cookie 信息，只是保存格式不同，读者可自行尝试
04 cookie = http.cookiejar.MozillaCookieJar(filename)
05 handler = urllib.request.HTTPCookieProcessor(cookie)
06 opener = urllib.request.build_opener(handler)
07 response = opener.open('http://www.baidu.com')
08 cookie.save(ignore_discard=True, ignore_expires=True)
```

保存结果如图 2.13 所示。

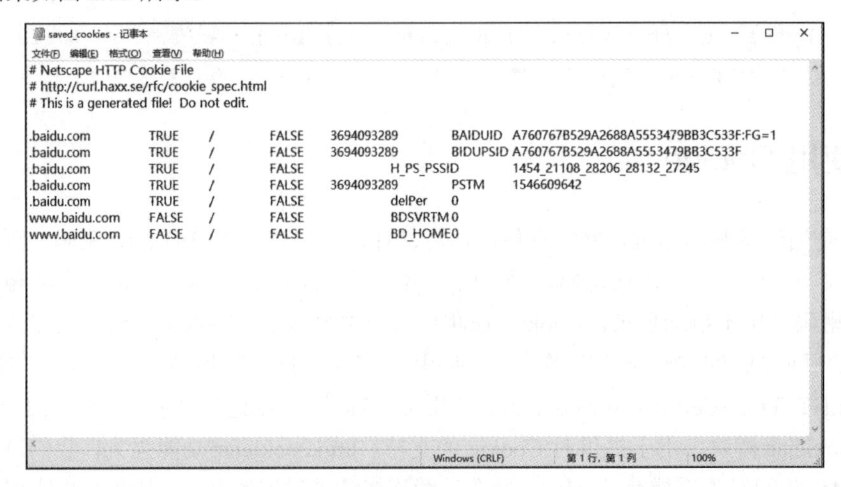

```
# Netscape HTTP Cookie File
# http://curl.haxx.se/rfc/cookie_spec.html
# This is a generated file!  Do not edit.

.baidu.com	TRUE	/	FALSE	3694093289	BAIDUID	A760767B529A2688A5553479BB3C533F:FG=1
.baidu.com	TRUE	/	FALSE	3694093289	BIDUPSID	A760767B529A2688A5553479BB3C533F
.baidu.com	TRUE	/	FALSE		H_PS_PSSID	1454_21108_28206_28132_27245
.baidu.com	TRUE	/	FALSE	3694093289	PSTM	1546609642
.baidu.com	TRUE	/	FALSE		delPer	0
www.baidu.com	FALSE	/	FALSE		BDSVRTM	0
www.baidu.com	FALSE	/	FALSE		BD_HOME	0
```

图 2.13 保存 Cookie

【示例 2-9】urllib 使用 Cookie

```
01 # 使用 Cookie
02 import http.cookiejar, urllib.request
03 cookie = http.cookiejar.MozillaCookieJar()
04 cookie.load('saved_cookies.txt', ignore_discard=True,
ignore_expires=True)
05 handler = urllib.request.HTTPCookieProcessor(cookie)
06 opener = urllib.request.build_opener(handler)
07 response = opener.open('http://www.baidu.com')
08 print(response.read().decode('utf-8'))
```

运行结果如图 2.14 所示。

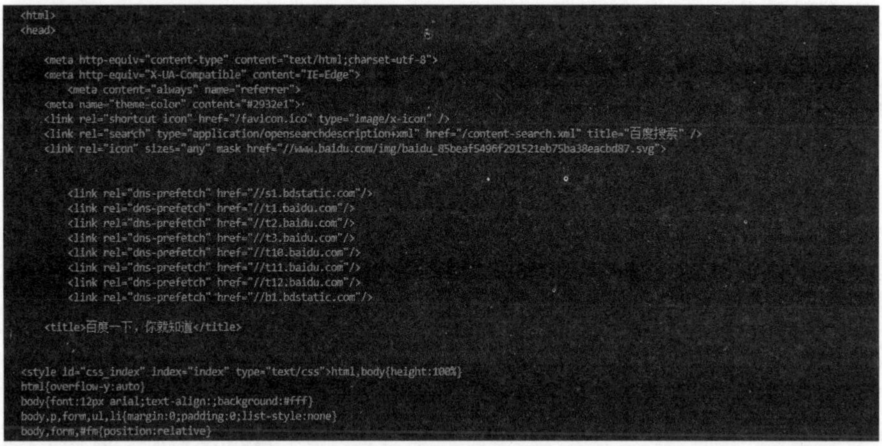

图 2.14 使用 Cookie

2.7 爬虫常用类库 2：更人性化的第三方库 requests

相比于 urllib，第三方库 requests 更加简单与人性化，是爬虫工作中常用的库，推荐大家安装。安装 requests 有两种方法：

- 通过 pip install requests 直接安装。
- 下载 Github 源码（https://github.com/requests/requests），运行 setup.py 进行安装。

下面看一下 requests 的基本使用方法。

【示例 2-10】requests 库的基本使用

```
01 import requests
02
03 r = requests.get(url='http://www.bing.com')
04 print(r.content)
```

可以看到，requests 发送 HTTP 请求非常简单，指定请求的 URL 之后，发送请求。其他几种 HTTP 请求方式为：

```
r = requests.post('http://httpbin.org/delete')
r = requests.put('http://httpbin.org/put', data = {'key':'value'})
r = requests.delete('http://httpbin.org/delete')
r = requests.head('http://httpbin.org/get')
r = requests.options('http://httpbin.org/get')
```

下面以 GET 与 POST 请求方式为例进行 requests 库的讲解。

2.7.1 发送请求

1. GET 请求

在基本用法中已经展示了如何发送 GET 请求。GET 方法一般用在查询中，查询都是带有参数进行查询，比如使用 Bing 搜索 requests，我们可以看到地址栏中的 URL 是这样的：

https://cn.bing.com/search?q=requests&qs=n&form=QBLH&sp=-1&pq=requests&sc=8-8&sk=&cvid=72590B4841C941B79094E826A134CC50

在开发人员工具中，我们可以看到查询时使用的参数，如图 2.15 所示。

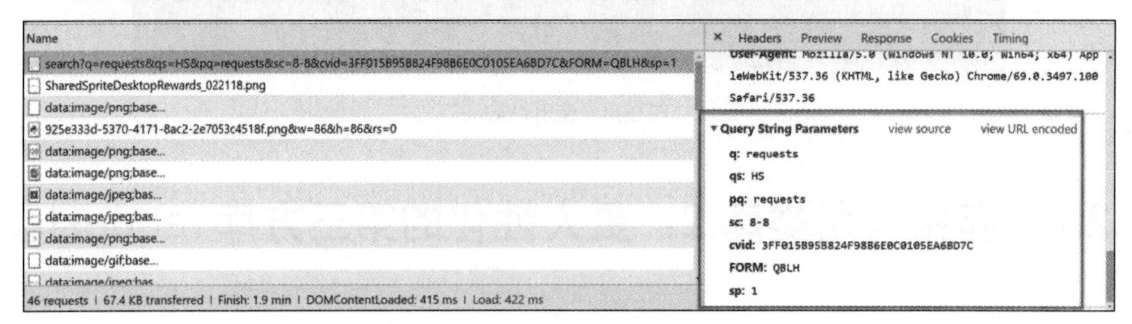

图 2.15 GET 请求参数

“q”对应的是查询数据，也就是“requests”，其他的则是浏览器自带的参数，代码如下：

【示例 2-11】requests 传递 URL 请求数据

```
01 import requests
02
03 payload = {
04     'q': 'requests',
05     'qs': 'HS',
06     'pq': 'requests',
07     'sc': '8-8',
08     'cvid': '3FF015B95B824F98B6E0C0105EA6BD7C',
09     'from': 'QBLH',
10     'sp': 1}
11 response = requests.get(url='http://www.bing.com', params=payload)
```

打印请求后的 URL：

http://cn.bing.com/search?q=requests&qs=HS&pq=requests&sc=8-8&cvid=3FF015B95B824F98B6E0C0105EA6BD7C&from=QBLH&sp=1

只需添加 params 参数即可。需要注意的是，请求参数中有值为 None 的键，是不会添加到 URL 的查询字符串中的。

2. POST 请求

POST 请求一般用在有数据交互的情形，比如注册账号、修改密码等。requests 传递 POST 很简单，只需将请求数据传递给 data 即可：

【示例 2-12】requests 传递表单数据

```
01 import requests
02
03 payload = {
04     'key1':'value1',
05     'key2':'value2'
06     }
07 response = requests.post(url='http://www.bing.com',data=payload)
```

有时请求的参数要求为 JSON 格式，则发送的时候使用 json 参数即可：

```
response = requests.post(url='http://www.bing.com', json=payload)
```

2.7.2 请求头

当我们打开一个网页时，浏览器要向网站服务器发送一个 HTTP 请求头，然后网站服务器根据 HTTP 请求头的内容生成当次请求的内容发送给浏览器。我们可以手动设定请求头的内容：

```
>>> import requests
>>> user_agent='Mozilla/5.0 (Windows NT 10.0; Win64; x64) AppleWebKit/537.36
(KHTML, like Gecko) Chrome/71.0.3578.98 Safari/537.36'
>>> headers = {'User-Agent':user_agent}
>>> response = requests.get('http://www.baidu.com',headers=headers)
```

2.7.3 响应内容

发送请求之后，会接收到服务器响应的内容，提取响应内容有两种形式：一种是文本形式；另一种是二进制形式。在文本形式中，requests 会根据响应头中的编码信息自行解码：

```
>>> import requests
>>> r = requests.get('http://blog.jobbole.com/all-posts/')
>>> r.encoding
'utf-8'
>>> r.text
'<!DOCTYPE html>… <link rel="apple-touch-icon"
href="http://jbcdn2.b0.upaiyun.com/2013/12/jobbole-blog-logo-mobile.png"
/>\r\n\t\r\n\t<!-- RSS, Atom & Pingbacks -->\r\n\t\t<link rel="alternate" title="
文章 &#8211; 伯乐在线 RSS Feed" href="http://blog.jobbole.com/feed/" />\r\n\...
>>> r.content
b'<!DOCTYPE html>…  <!-- RSS, Atom & Pingbacks -->\r\n\t\t<link
```

```
rel="alternate" title="\xe6\x96\x87\xe7\xab\xa0 &#8211;
\xe4\xbc\xaf\xe4\xb9\x90\xe5\x9c\xa8\xe7\xba\xbf RSS Feed"
href="http://blog.jobbole.com/feed/" …
```

当返回内容为 JSON 格式时，可使用 requests 内置的 JSON 解码器进行数据处理：

```
    >>> import requests

    >>> r=requests.get('https://segmentfault.com/api/live/recommend?')
    >>> r.json()
    {'status': 0, 'data': [{'id': '1500000015237807', 'url':
'/l/1500000015237807', 'title': '浏览器工作原理及在网页性能优化中的应用', 'startWord':
'2018-06-11 周一', 'joinedUsers': '131', 'averageScore': '4.7', 'isFinished': 1,
'price': '9.90', …}
```

2.7.4 响应状态码

服务器返回响应内容时，也包含相应的响应码：

```
    >>> import requests

    >>> r = requests.get('http://bing.com')
    >>> r.status_code
    200
```

requests 也内置了状态码查询对象：

```
    >>> r = requests.get('http://bing.com')
    >>> r.status_code
    200
    >>> r.status_code == requests.codes.ok
    True
```

如果 HTTP 请求发生错误，比如 4**客户端错误、5**服务器错误，则可以使用 raise_for_status() 抛出异常：

```
    >>> r=requests.get('https://cn.bing.com/no-this-page')
    >>> r.status_code
    404
    >>> r.raise_for_status()
    Traceback (most recent call last):
      File "<stdin>", line 1, in <module>
      File "C:\Users\test\AppData\Local\Programs\Python\Python36\lib\
site-packages\requests\models.py", line 939, in raise_for_status
        raise HTTPError(http_error_msg, response=self)
    requests.exceptions.HTTPError: 404 Client Error: Not Found for url:
https://www4.bing.com/no-this-page
```

2.7.5　cookies 参数

requests 操作 Cookie 很简单，只需指定 cookies 参数即可。

Cookie 内容需要以 dict 形式传递。

在响应中，可以指定字段获取 Cookie 值：

```
>>> import requests

>>> r = requests.get('https://www.csdn.net/')
>>> r.cookies['uuid_tt_dd']
'1489347943734346625_20181014'
```

要在请求的时候携带 Cookie，只需指定 cookies 参数即可：

```
>>> my_cookies = {'test_cookie':'this is a test'}
>>> r = requests.get('http://www.bing.com',cookies=my_cookies)
```

2.7.6　重定向与请求历史

重定向是指将网络请求重新转向其他位置，例如网站迁移，访问原网址时自动转到新的网址，或者其他原因跳转到新的网址。默认情况下，除了 head 方法外，requests 会自动处理所有的重定向，也可以手动设定 allow_redirects 参数为 True 或者 False 来开启或禁止重定向，例如 r=requests.get('http://www.z.cn',allow_redirects=False)。

如果允许重定向，可以使用响应对象的 history 方法来查看跳转信息。查看如下示例，Github 将所有的 HTTP 请求重定向到 HTTPS：

```
>>> import requests
>>>
>>>r = requests.get('http://github.com')
>>> r.url
'https://github.com/'
>>>
>>> r.status_code
200
>>>
>>> r.history
[<Response [301]>]
```

2.7.7 超时

可以指定 timeout 参数来设定等待网站响应的时间（单位：秒），超过设定时间之后停止等待。

```
>>>import requests
>>> requests.get('http://www.bing.com', timeout=0.001)
….
requests.exceptions.ConnectTimeout: HTTPConnectionPool(host='www.bing.com', port=80): Max retries exceeded with url: / (Caused by ConnectTimeoutError (<urllib3.connection.HTTPConnection object at 0x00000266B6367208>, 'Connection to www.bing.com timed out. (connect timeout=0.001)'))
```

2.7.8 设置代理

当某些网站对访问 IP 有限制（如地域限制或次数限制）时，我们可以使用代理来绕过限制。requests 使用代理只需指定 proxies 参数即可：

```
>>>import requests
>>>
>>> proxie = {
...     "http":"http://10.10.1.10:3128",
...     "https":"http://10.10.1.10:1080",
... }
>>>
>>>requests.get("http://example.org", proxies=proxies)
```

如果代理需要使用验证，可以使用 http://user:password@host/语法：

```
>>>proxies = {
...      "http": "http://user:pass@10.10.1.10:3128/",
}
```

2.7.9 会话对象

会话对象是一种高级的用法，可以跨请求保持某些参数，比如可以在同一个 Session 实例之间保存 Cookie，像浏览器一样，我们并不需要每次请求指定 Cookie，Session 会自动在后续的请求中添加获取的 Cookie，这种处理方式在同一站点连续请求中特别方便：

```
>>>import requests
>>>
>>> s = requests.Session()
>>>
>>> s.get('http://httpbin.org/cookies/set/sessioncookie/123456789')
>>> r = s.get("http://httpbin.org/cookies")
```

```
>>>
>>> print(r.text)
{
  "cookies": {
    "sessioncookie": "123456789"
  }
}
```

但需要注意的是，即使使用了 Session 对象，方法级别的参数也不会跨请求保持，例如在发送请求时，我们手动指定的 cookies 参数，在下一次发送请求时并不会保存使用。下面的例子在第一次请求时指定的参数，在第二次请求时失效：

```
>>> import requests

>>>s = requests.Session()
>>>
>>>r = s.get('http://httpbin.org/cookies', cookies={'first-request-cookies':
'first'})
>>>print(r.text)
{
  "cookies": {
    "first-request-cookies": "first"
  }
}
>>>
>>>r = s.get('http://httpbin.org/cookies')
>>>print(r.text)
{
  "cookies": {}
}
```

2.8 爬虫常用类库 3：元素提取利器 BeautifulSoup

BeautifulSoup 是一个可以从 HTML 或 XML 文件中提取数据的 Python 库，可以很方便地进行文档的导航、查找或修改。在爬虫工作中，常用的是查找功能。

2.8.1 安装 BeautifulSoup

如果使用的是新版的 Debian 或 Ubuntu，那么可以通过系统的软件包管理来安装：

```
apt-get install Python-bs4
```

BeautifulSoup 4 通过 PyPi 发布，如果无法使用系统包管理安装，那么可以通过 easy_install 或 pip 来安装。包的名字是 beautifulsoup4，这个包兼容 Python 2 和 Python 3。

```
easy_install beautifulsoup4
pip install beautifulsoup4
```

2.8.2 安装解析器

BeautifulSoup 支持 Python 标准库中的 HTML 解析器，还支持一些第三方的解析器，其中一个是 lxml。lxml 解析速度很快，在后面的爬虫开发中我们主要使用此解析器。根据操作系统不同，可以选择下列方法来安装 lxml：

```
apt-get install Python-lxml
easy_install lxml
pip install lxml
```

在 Windows 系统下，如果在安装过程中提示其他错误信息，这几种方法都无法安装，那么可以下载 lxml 文件安装，在网站 https://www.lfd.uci.edu/~gohlke/pythonlibs/#lxml 提供了很多第三方安装包，根据 Python 版本及操作系统选择对应的文件下载，如图 2.16 所示。

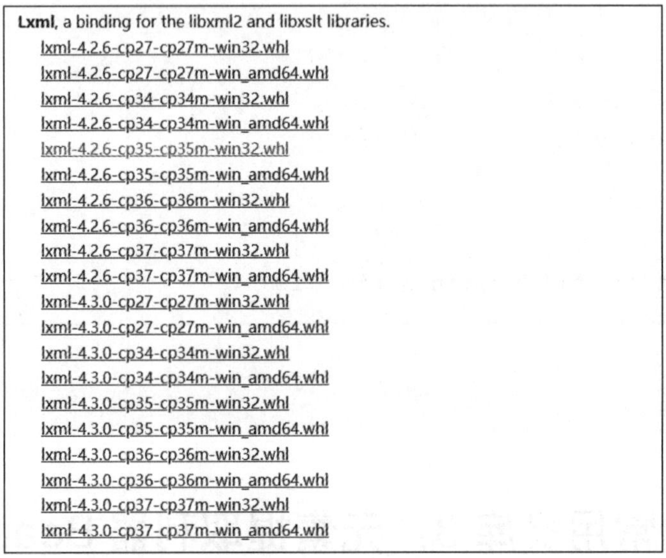

Lxml, a binding for the libxml2 and libxslt libraries.

- lxml-4.2.6-cp27-cp27m-win32.whl
- lxml-4.2.6-cp27-cp27m-win_amd64.whl
- lxml-4.2.6-cp34-cp34m-win32.whl
- lxml-4.2.6-cp34-cp34m-win_amd64.whl
- lxml-4.2.6-cp35-cp35m-win32.whl
- lxml-4.2.6-cp35-cp35m-win_amd64.whl
- lxml-4.2.6-cp36-cp36m-win32.whl
- lxml-4.2.6-cp36-cp36m-win_amd64.whl
- lxml-4.2.6-cp37-cp37m-win32.whl
- lxml-4.2.6-cp37-cp37m-win_amd64.whl
- lxml-4.3.0-cp27-cp27m-win32.whl
- lxml-4.3.0-cp27-cp27m-win_amd64.whl
- lxml-4.3.0-cp34-cp34m-win32.whl
- lxml-4.3.0-cp34-cp34m-win_amd64.whl
- lxml-4.3.0-cp35-cp35m-win32.whl
- lxml-4.3.0-cp35-cp35m-win_amd64.whl
- lxml-4.3.0-cp36-cp36m-win32.whl
- lxml-4.3.0-cp36-cp36m-win_amd64.whl
- lxml-4.3.0-cp37-cp37m-win32.whl
- lxml-4.3.0-cp37-cp37m-win_amd64.whl

图 2.16 手动下载 lxml 包

下载完成后，执行如下命令进行安装：

```
pip install lxml-4.3.0-cp37-cp37m-win_amd64.whl
```

另一个可供选择的解析器是纯 Python 实现的 html5lib。html5lib 的解析方式与浏览器相同，可以选择下列方法来安装 html5lib：

```
apt-get install Python-html5lib
easy_install html5lib
pip install html5lib
```

表 2-22 列出了主要的解析器以及它们的优缺点。

表2-22　主要解析器对比

解析器	使用方法	优势	劣势
Python 标准库	BeautifulSoup(markup, "html.parser")	• Python 的内置标准库 • 执行速度适中 • 文档容错能力强	Python 2.7.3 或 3.2.2 前的版本中，文档容错能力差
lxml HTML 解析器	BeautifulSoup(markup, "lxml")	• 速度快 • 文档容错能力强	需要安装 C 语言库
lxml XML 解析器	BeautifulSoup(markup, ["lxml", "xml"]) BeautifulSoup(markup, "xml")	• 速度快 • 唯一支持 XML 的解析器	需要安装 C 语言库
html5lib	BeautifulSoup(markup, "html5lib")	• 最好的容错性 • 以浏览器的方式解析文档 • 生成 HTML 5 格式的文档	速度慢，不依赖外部扩展

2.8.3 BeautifulSoup 使用方法

有以下包含 HTML 代码的字符串，后面的示例都以此字符串为例进行操作的：

```
01 html_doc = """
02 <html><head><title>The Dormouse's story</title></head>
03 <body>
04 <p class="title"><b>The Dormouse's story</b></p>
05
06 <p class="story">Once upon a time there were three little sisters;
          and their names were
07 <a href="http://example.com/elsie" class="sister" id="link1">Elsie</a>,
08 <a href="http://example.com/lacie" class="sister" id="link2">
         Lacie</a> and
09 <a href="http://example.com/tillie" class="sister" id="link3">
         Tillie</a>;
10 and they lived at the bottom of a well.</p>
11
12 <p class="story">...</p>
13 </body>
14 </html>
15 """
```

进行文档解析，需要先创建 BeautifulSoup 对象，有两种方式可供选择。

（1）直接通过字符串进行创建

```
from bs4 import BeautifulSoup
soup = BeautifulSoup(html_doc,'lxml')
```

BeautifulSoup 将使用指定的解析器来解析文档，若未指定，则自动选择合适的解析器进行解析，这里我们指定为 lxml。

（2）通过打开文件来创建

假设我们将 HTML 字符串中的 HTML 代码保存为 index.html 文件，可以这样创建对象：

```
01 from bs4 import BeautifulSoup
02 soup = BeautifulSoup(open('index.html'))
```

BeautifulSoup 对象可以按照标准缩进格式结构输出：

```
01 print(soup.prettify())
02
03 <html>
04  <head>
05   <title>
06    The Dormouse's story
07   </title>
08  </head>
09  <body>
10   <p class="title">
11    <b>
12     The Dormouse's story
13    </b>
14   </p>
15   <p class="story">
16    Once upon a time there were three little sisters; and their names were
17    <a class="sister" href="http://example.com/elsie" id="link1">
18     Elsie
19    </a>
20    ,
21    <a class="sister" href="http://example.com/lacie" id="link2">
22     Lacie
23    </a>
24    and
25    <a class="sister" href="http://example.com/tillie" id="link3">
26     Tillie
27    </a>
28    ;
29    and they lived at the bottom of a well.
30   </p>
31   <p class="story">
32    ...
33   </p>
34  </body>
35 </html>
```

2.8.4　BeautifulSoup 对象

BeautifulSoup 将复杂 HTML 文档转换成一个复杂的树形结构，每个节点都是 Python 对象，所有对象可以归纳为 4 种：

- Tag
- NavigableString
- BeautifulSoup
- Comment

下面将对这 4 种对象做详细介绍。

1. Tag

Tag 对象与 XML 或 HTML 原生文档中的标签相同，如<title>、<a>、<p>等，这些标签及其内容就构成了一个 Tag 对象。Tag 对象使用 BeautifulSoup 对象加对应标签名来提取，如：

```
# 提取 title 对象
title = soup.title
# 提取 a 对象
a = soup.a
```

Tag 对象有很多属性和方法，其中最重要和常用的两个属性是 name 和 attributes。每个 tag 都有自己的名字，可以通过.name 来获取：

```
print(title.name)
print(a.name)
```

输出：

```
title
a
```

Tag 也可以改变 name 值，如果做了更改，就会影响所有通过当前 BeautifulSoup 对象生成的 HTML 文档。

```
# 抽取 title 对象
tag = soup.title
print("tag 内容:")
print(tag)
print("tag 名称:" + tag.name)
# 更改 title 的 name 值
tag.name = 'newtitle'
print("修改后 tag 内容:")
print(tag)
print("修改后 tag 名称:" + tag.name)
```

输出结果：

```
tag 内容:
<title>The Dormouse's story</title>
tag 名称:title
修改后 tag 内容:
<newtitle>The Dormouse's story</newtitle>
修改后 tag 名称:newtitle
```

可以看到，原 title 标签已经改为 newtitle。

每个 Tag 可能包含多个属性，如 Tag <p class="title"><b>The Dormouse's story</b></p>中有一个 class 属性，值为“title”。属性的操作方式与字典相同：

```
tag = soup.p
print(tag)
print(tag['class'])
```

输出结果：

```
<p class="title"><b>The Dormouse's story</b></p>
['title']
```

也可以直接“点”取属性，比如使用.attrs 来获取属性：

```
print(tag.attrs)
```

输出结果：

```
{'class': ['title']}
```

Tag 中的属性可以增加、修改、删除：

```
tag = soup.a
print("修改前 tag 内容与属性:")
print(tag)
print(tag.attrs)
# 增加 name 属性
tag['name'] = 'link-name'
# 修改 class 属性
tag['class'] = 'brother'
# 删除 id 属性
del tag['id']
print("修改后 tag 内容与属性: ")
print(tag)
print(tag.attrs)
```

输出结果为：

```
修改前 tag 内容与属性:
<a class="sister" href="http://example.com/elsie" id="link1">Elsie</a>
{'href': 'http://example.com/elsie', 'class': ['sister'], 'id': 'link1'}
修改后 tag 内容与属性:
<a class="brother" href="http://example.com/elsie"
name="link-name">Elsie</a>
```

```
{'href': 'http://example.com/elsie', 'class': 'brother', 'name':
'link-name'}
```

有些情况下一个属性的值可能不止一个，如<p class="semantic ui"></p>称为多值属性。在 BeautifulSoup 中，多值属性的返回类型是一个 list：

```
from bs4 import BeautifulSoup

soup = BeautifulSoup('<p class="semantic ui"></p>','lxml')
print(soup.p['class'])
```

输出结果：

```
['semantic', 'ui']
```

如果某个属性有多个值，但该属性在任何 HTML 版本中都没有定义为多值属性，那么将返回一个字符串而不是 list：

```
from bs4 import BeautifulSoup

soup = BeautifulSoup('<p no-definition="semantic ui"></p>','lxml')
print(soup.p[' no-definition'])
```

输出结果：

```
semantic ui
```

2. NavigableString

BeautifulSoup 使用 NavigableString 类来包装 Tag 中的字符串，可以通过.string 提取 Tag 中的字符串数据：

```
soup = BeautifulSoup('<p class="page">This is a test</p>','lxml')
tag = soup.p
print(tag.string)
```

输出结果：

```
This is a test
```

Tag 中包含的字符串不能编辑，但是可以用 replace_with()方法替换成其他的字符串：

```
soup = BeautifulSoup('<p class="page">This is a test</p>','lxml')
tag = soup.p
print(tag)
tag.string.replace_with("This is another test")
print(tag)
```

输出结果：

```
<p class="page">This is a test</p>
<p class="page">This is another test</p>
```

3. BeautifulSoup

BeautifulSoup 对象表示的是一个文档的全部内容。大部分时候，可以把它当作一个 Tag 对象。因为 BeautifulSoup 对象并不是真正的 HTML 或 XML 的标签，所以它没有 name 和 attribute 属性。但有时查看它的.name 属性是很有用的，所以 BeautifulSoup 对象被赋予了一个值为“[document]”的特殊属性.name。

```
print(soup.name)
```

输出结果：

```
[document]
```

4. Comment

Tag、NavigableString、BeautifulSoup 几乎覆盖了 HTML 和 XML 中的所有内容，但是还有一些特殊对象，需要注意的内容是文档的注释部分：

```
01 from bs4 import BeautifulSoup
02
03 html_doc = "<p><!--This is a comment --></p>"
04 soup = BeautifulSoup(html_doc,'lxml')
05 comment = soup.p.string
06 print(type(comment))
07 print(comment)
```

输出结果：

```
<class 'bs4.element.Comment'>
This is a comment
```

可以看到，虽然可以使用.string 提取文本数据，但是输出的类型是一个 Comment 类型，这就会出现一个问题，当我们没有判断数据类型时，会提取到垃圾数据，处理方法是加上一个类型判断：

```
01 from bs4 import BeautifulSoup
02 import bs4
03
04 html_doc = "<p><!--This is a comment --></p>"
05 soup = BeautifulSoup(html_doc,'lxml')
06 comment = soup.p.string
07 if type(comment) == bs4.element.Comment:
08     print("comment string,drop it")
```

输出结果：

```
comment string,drop it
```

2.8.5 遍历文档树

HTML 与 XML 文档都是由一个个标签组成的，这些标签可以看作整个文档树的节点。像树杈一样，一个节点既有上级节点，又包含次级节点；次级节点相对于该节点称为子节点，上级节点称为父节点，与该节点同级并列的称为兄弟节点，按照节点顺序有前后节点。下面将分别介绍。

我们仍使用以下包含 HTML 代码的字符串作为操作示例进行讲解：

```
01 html_doc = """
02 <html>
01 <head><title>The Dormouse's story</title></head>
02 <body>
03     <p class="title"><b>The Dormouse's story</b></p>
04     <p class="story">Once upon a time there were three little sisters;
                 and their names were
05         <a href="http://example.com/elsie" class="sister" id="link1">
                 Elsie</a>,
06         <a href="http://example.com/lacie" class="sister" id="link2">
                 Lacie</a> and
07         <a href="http://example.com/tillie" class="sister" id="link3">
                 Tillie</a>;
08         and they lived at the bottom of a well.</p>
09     <p class="story">...</p>
10 </body>
11 </html>
12 """
13
14 from bs4 import BeautifulSoup
15 soup = BeautifulSoup(html_doc)
```

1. 子节点

一个 Tag 可以包含其他的 Tag，称为该 Tag 的子节点。在前面介绍 Tag 对象时，我们知道获取一个标签最简单的方法是使用 Tag 的 name，如获取<head>只要用 soup.head 即可。这个方法可以嵌套调用，如想获取<body>中的<p>标签，可以这样：

```
soup.body.p
```

获取到：

```
<p class="title"><b>The Dormouse's story</b></p>
```

需要注意的是，如果 body 下有多个<p>标签，通过这种方式，只能获取到第一个<p>标签。Tag 的.contents 可以将所有子节点以列表的方式输出：

```
tag = soup.body
print(tag.contents)
```

输出结果：

```
['\n',
<p class="title"><b>The Dormouse's story</b></p>,
'\n',
 <p class="story">Once upon a time there were three little sisters; and their
names were
         <a class="sister" href="http://example.com/elsie" id="link1">
Elsie</a>,
         <a class="sister" href="http://example.com/lacie" id="link2">
Lacie</a> and
         <a class="sister" href="http://example.com/tillie" id="link3">
Tillie</a>;
         and they lived at the bottom of a well.</p>,
'\n',
<p class="story">...</p>,
'\n']
```

我们可以获取该列表的大小，并通过索引获取里面的值：

```
tag = soup.body
print(len(tag.contents))
print(tag.contents[5])
```

输出结果：

```
7
<p class="story">...</p>
```

字符串没有.contents 属性，因为字符串没有子节点。

Tag 的.children 属性可以返回一个生成器，对子节点进行循环：

```
tag = soup.body
for i in tag.children:
   print(i)
```

输出结果：

```
<p class="title"><b>The Dormouse's story</b></p>


<p class="story">Once upon a time there were three little sisters; and their
names were
         <a class="sister" href="http://example.com/elsie" id="link1">
Elsie</a>,
         <a class="sister" href="http://example.com/lacie" id="link2">
Lacie</a> and
```

```
        <a class="sister" href="http://example.com/tillie" id="link3">
Tillie</a>;
        and they lived at the bottom of a well.</p>

    <p class="story">...</p>
```

如果Tag仅包含一个字符串，或者仅包含一个子节点，那么可使用.string来获取这个节点的字符串或者子节点的字符串：

```
# <head><title>The Dormouse's story</title></head>
print(soup.title.string)
print(soup.head.string)
```

输出结果：

```
The Dormouse's story
The Dormouse's story
```

当Tag中包含多个子节点时，子节点无法直接使用.string来获取内容，因为无法确定.string方法该调用哪个子节点的内容，导致输出结果为None：

```
print(soup.body.string)
```

输出结果：

```
None
```

多个节点的情况可以使用.strings属性循环获取内容：

```
for string in soup.strings:
print(repr(string))
```

输出结果：

```
'\n'
"The Dormouse's story"
'\n'
'\n'
"The Dormouse's story"
'\n'
'Once upon a time there were three little sisters; and their names were\n        '
'Elsie'
',\n        '
'Lacie'
' and\n        '
'Tillie'
';\n        and they lived at the bottom of a well.'
'\n'
'...'
'\n'
'\n'
```

BeautifulSoup 对象也看作一个节点，使用.strings 输出全部内容。可以使用 stripped_string 去除换行和段首、段末的空白内容：

```
for string in soup.strings:
    print(repr(string))
```

输出结果：

```
"The Dormouse's story"
"The Dormouse's story"
'Once upon a time there were three little sisters; and their names were'
'Elsie'
','
'Lacie'
'and'
'Tillie'
';\n       and they lived at the bottom of a well.'
'...'
```

2. 父节点

每个 Tag 或字符串都有父节点，HTML 文档顶层节点<html>的父节点是 BeautifulSoup 对象。可以通过.parent 属性来获取某个元素的父节点：

```
# <head><title>The Dormouse's story</title></head>
title_tag = soup.title
print(title_tag)
print(title_tag.parent)
```

输出结果：

```
<title>The Dormouse's story</title>
<head><title>The Dormouse's story</title></head>
```

通过元素的.parents 属性可以递归得到元素的所有父节点，以<body>标签中第一个<a>标签为例：

```
# <a href="http://example.com/elsie" class="sister" id="link1">Elsie</a>
for parent in soup.a.parents:
    if parent is None:
        print(parent)
    else:
        print(parent.name)
```

输出结果：

```
p
body
html
[document]
```

3．兄弟节点

兄弟节点属于同一级，比如在示例文档中，<body>标签下的 3 个<p>标签互为兄弟节点，第二个<p>标签下的 3 个<a>标签互为兄弟节点，都是同一个元素的子节点。

可以使用.next_sibling 和.previous_sibling 属性来查询兄弟节点：

```
print(soup.a.next_sibling)
print(soup.a.next_sibling.next_sibling)
print(soup.body.previous_sibling)
print(soup.body.previous_sibling.previous_sibling)
```

输出结果：

```
,
<a class="sister" href="http://example.com/lacie" id="link2">Lacie</a>

<head><title>The Dormouse's story</title></head>
None
```

输出结果中第二行的<a>标签紧跟着的兄弟节点为“，”。再次明确一点，左右的字符串、换行、空白都可以视作一个节点。代码中第二行 print 语句打印出<a>标签的第二个兄弟标签，也就是“，”的下一个兄弟标签。代码中第三行 print 语句打印出<body>标签的前一个兄弟标签，是一个换行，下一条输出语句打印出<body>第二个前兄弟标签，也就是<head>。代码中最后一条输出语句打印出<html>标签的下一个兄弟节点，因为<html>标签没有兄弟节点，所以为 None。

通过.next_siblings 和.previous_siblings 属性可以对兄弟节点进行迭代输出：

```
for i in soup.a.next_siblings:
    print(repr(i))
```

输出结果：

```
',\n        '
<a class="sister" href="http://example.com/lacie" id="link2">Lacie</a>
' and\n        '
<a class="sister" href="http://example.com/tillie" id="link3">Tillie</a>
';\n        and they lived at the bottom of a well.'
```

4．前后节点

前后节点可以理解为该节点前后位置的节点，注意与兄弟节点按等级划分不同，如<html><head><title>The Dormouse's story</title></head>，<head>下一个节点就是<title>。

可以使用.next_element 和.previous_element 属性来访问前后节点：

```
# <head><title>The Dormouse's story</title></head>
print(soup.head.next_element)
print(soup.title.previous_element)
```

输出结果：

```
<title>The Dormouse's story</title>
<head><title>The Dormouse's story</title></head>
```

通过.next_elements 和.previous_elements 的迭代器就可以向前或向后访问文档的解析内容，就好像文档正在被解析一样：

```
soup = BeautifulSoup("<head><title>The Dormouse's story</title></head>",
'lxml')
for element in soup.head.next_elements:
    print(repr(element))
```

输出结果：

```
<title>The Dormouse's story</title>
"The Dormouse's story"
```

2.8.6 搜索文档树

获取到网页数据之后，接下来的核心工作就是提取、采集数据。BeautifulSoup 提供了很多搜索数据的方法，本节着重介绍两个方法：find()和 find_all()，其他方法读者可自行学习。

我们的操作示例仍然是：

```
01 html_doc = """
02 <html><head><title>The Dormouse's story</title></head>
03 <body>
04 <p class="title"><b>The Dormouse's story</b></p>
05
06 <p class="story">Once upon a time there were three little sisters;
           and their names were
07 <a href="http://example.com/elsie" class="sister" id="link1">Elsie</a>,
08 <a href="http://example.com/lacie" class="sister" id="link2">
           Lacie</a> and
09 <a href="http://example.com/tillie" class="sister" id="link3">
           Tillie</a>;
10 and they lived at the bottom of a well.</p>
11
12 <p class="story">...</p>
13 """
14 from bs4 import BeautifulSoup
15 soup = BeautifulSoup(html_doc,'lxml')
```

先来看看这两个搜索方法支持的筛选类型，这些筛选类型可以使用字符串、Tag 名称、属性，甚至方法，并且可以混合使用，看下面的例子：

1. 字符串筛选

传入字符串是最简单的筛选方式，以下操作将筛选出所有的<b>标签：

```
print(soup.find_all('b'))
```

输出结果：

```
[<b>The Dormouse's story</b>]
```

2. 正则表达式

如果传入一个正则表达式对象，BeautifulSoup 就会通过正则表达式的 search()方法来匹配内容，如下将匹配<b>和<body>标签：

```
01 import re
02 soup = BeautifulSoup(html_doc,'lxml')
03 for tag in soup.find_all(re.compile("^b")):
04     print(tag.name)
```

输出结果：

```
body
b
```

3. 列表

传入一个列表时，将会匹配列表中的任意一个标签元素：

```
for tag in soup.find_all(["a","b"]):
    print(tag)
```

输出结果：

```
<b>The Dormouse's story</b>
<a class="sister" href="http://example.com/elsie" id="link1">Elsie</a>
<a class="sister" href="http://example.com/lacie" id="link2">Lacie</a>
<a class="sister" href="http://example.com/tillie" id="link3">Tillie</a>
```

4. True

传入 True 时，会返回所有的节点，但不包括字符串节点：

```
[<b>The Dormouse's story</b>]
for tag in soup.find_all(True):
    print(tag.name)
```

输出结果：

```
html
head
title
body
```

```
p
b
p
a
a
a
p
```

5. 自定义方法

当没有合适的筛选参数可供选择时，可以自定义一个筛选方法。该方法只能接收一个元素作为参数，若元素满足方法中的条件，则返回 True，否则返回 False。下例中，筛选出既含有 class 属性又含有 id 属性的元素：

```
<b>The Dormouse's story</b>
def has_class_and_id(tag):
    return tag.has_attr("class") and tag.has_attr('id')
for tag in soup.find_all(has_class_and_id):
    print(tag)
```

输出结果：

```
<a class="sister" href="http://example.com/elsie" id="link1">Elsie</a>
<a class="sister" href="http://example.com/lacie" id="link2">Lacie</a>
<a class="sister" href="http://example.com/tillie" id="link3">Tillie</a>
```

6. find_all()方法

find_all()方法原型：

```
find_all(name, attrs, recursive, string, limit, **kwargs)
```

参数说明：

（1）name 参数，可以查找所有名字为 name 的 Tag，字符串对象会被自动忽略掉。类型可以为上述介绍的筛选参数的任意类型。

（2）kwargs 参数，任何未被识别为 Tag 的参数都将作为 Tag 属性来筛选数据，例如传递一个 id 参数，BeautifulSoup 将会搜索每个 Tag 的 id 属性：

```
print(soup.find_all(id="link1"))
```

输出结果：

```
[<a class="sister" href="http://example.com/elsie" id="link1">Elsie</a>]
```

搜索指定名字的属性时，不仅可以使用字符串作为参数值，正则表达式、列表、True 都可以使用。

使用正则表达式搜索：

```
print(soup.find_all(href=re.compile("elsie")))
```

输出结果：

```
[<a class="sister" href="http://example.com/elsie" id="link1">Elsie</a>]
```

使用列表作为属性值搜索，会查找对应于列表中每一个值的 Tag：

```
print(soup.find_all(id=["link1","link3"]))
```

输出结果：

```
[<a class="sister" href="http://example.com/elsie" id="link1">Elsie</a>,
<a class="sister" href="http://example.com/tillie" id="link3">Tillie</a>]
```

使用 True 作为属性值，只要包括该属性的 Tag 都会被搜索到：

```
print(soup.find_all(id=True))
```

输出结果：

```
[<a class="sister" href="http://example.com/elsie" id="link1">Elsie</a>,
<a class="sister" href="http://example.com/lacie" id="link2">Lacie</a>,
<a class="sister" href="http://example.com/tillie" id="link3">Tillie</a>]
```

仍然有些属性在搜索的时候不能使用，比如 HTML 5 中的 data-*属性：

```
data_soup = BeautifulSoup('<div data-foo="value">foo!</div>')
print(data_soup.find_all(data-foo="value"))
```

这样会报出一个错误：

```
SyntaxError: keyword can't be an expression
```

说明 data-foo 不能作为属性被直接搜索，但是我们可以通过 attr 参数定义一个字典来搜索这类特殊属性：

```
data_soup = BeautifulSoup('<div data-foo="value">foo!</div>')
print(data_soup.find_all(attrs={"data-foo":"value"}))
```

输出结果：

```
[<div data-foo="value">foo!</div>]
```

还有一个需要特别注意的地方是，使用 class 属性搜索时，由于和 Python 中的关键字 class 冲突，因此在使用该属性时需要写作 class_，注意下画线：

```
soup = BeautifulSoup(html_doc,'lxml')
print(soup.find_all('a',class_='sister'))
```

输出结果：

```
[<a class="sister" href="http://example.com/elsie" id="link1">Elsie</a>,
<a class="sister" href="http://example.com/lacie" id="link2">Lacie</a>,
<a class="sister"href="http://example.com/tillie" id="link3">Tillie</a>]
```

（3）text 参数，可以搜索文档中的字符串内容，还可以查找 Tag。与 name 参数的可选值一样，text 参数接收字符串、正则表达式、列表、True：

```
print(soup.find_all(text="Elsie"))
print(soup.find_all(text=['Elsie','Lacie','Tillie']))
print(soup.find_all(text=re.compile('Dormouse')))
print(soup.find_all("a", text="Elsie"))
```

输出结果：

```
['Elsie']
['Elsie', 'Lacie', 'Tillie']
["The Dormouse's story", "The Dormouse's story"]
[<a class="sister" href="http://example.com/elsie" id="link1">Elsie</a>]
```

（4）limit 参数，限制返回的结果数。如下例，符合条件的结果有 3 个，通过 limit 参数限制返回两个：

```
print(soup.find_all('a',limit=2))
```

输出结果：

```
[<a class="sister" href="http://example.com/elsie" id="link1">Elsie</a>,
<a class="sister" href="http://example.com/lacie" id="link2">Lacie</a>]
```

（5）recursive 参数，调用 Tag 的 find_all()方法时，BeautifulSoup 会检索当前 Tag 的所有子孙节点，如果只想搜索 Tag 的直接子节点，则可以使用参数 recursive=False。

7. find()方法

与 find_all()唯一的不同点是，find_all()返回的是满足条件的所有结果的列表，而 find()只返回一条结果。

除了 find_all()与 find()方法外，BeautifulSoup 中还有 10 个用于搜索的 API。它们中的 5 个用的是与 find_all()相同的搜索参数，另外 5 个与 find()方法的搜索参数类似，区别仅仅是文档搜索的部分不同，见表 2-23。

表 2-23　find()方法的使用

方法	说明
find_parents(name , attrs , recursive , text , **kwargs) find_parent(name , attrs , recursive , text , **kwargs)	用来搜索当前节点的父辈节点，find_parent()查找符合条件的第一个结果，find_parents()返回符合条件的结果列表
find_next_siblings(name , attrs , recursive , text , **kwargs) find_next_sibling(name , attrs , recursive , text , **kwargs)	通过.next_siblings 属性对当前 tag 的所有后面的兄弟 tag 节点进行迭代，find_next_siblings()方法返回所有符合条件的后面的兄弟节点列表，find_next_sibling()只返回符合条件的后面的第一个 tag 节点

（续表）

方法	说明
find_previous_siblings(name,attrs, recursive , text , **kwargs) find_previous_sibling(name, attrs , recursive , text , **kwargs)	通过.previous_siblings 属性对当前 tag 的前面的兄弟 tag 节点进行迭代，find_previous_siblings()方法返回所有符合条件的前面的兄弟节点，find_previous_sibling()方法返回第一个符合条件的前面的兄弟节点
find_all_next(name , attrs , recursive , text , **kwargs) find_next(name , attrs , recursive , text , **kwargs)	通过.next_elements 属性对当前 tag 之后的 tag 和字符串进行迭代，find_all_next()方法返回所有符合条件的节点，find_next()方法返回第一个符合条件的节点
find_all_previous(name , attrs , recursive , text , **kwargs) find_previous(name , attrs , recursive , text , **kwargs)	通过.previous_elements 属性对当前节点前面的 tag 和字符串进行迭代，find_all_previous()方法返回所有符合条件的节点，find_previous()方法返回第一个符合条件的节点

2.8.7　BeautifulSoup 中的 CSS 选择器

BeautifulSoup 支持大部分的 CSS 选择器，在 Tag 或 BeautifulSoup 对象的.select()方法中传入对应 CSS 语法的字符串参数即可找到目标 Tag：

（1）通过 Tag 标签名查找

```
# 通过标签名直接查找
print(soup.select("title"))
# 通过标签逐层查找
print(soup.select("html title"))
# 查找某个 Tag 标签下的直接子标签
print(soup.select("p > a"))
```

输出结果：

```
[<title>The Dormouse's story</title>]
[<title>The Dormouse's story</title>]
[<a class="sister" href="http://example.com/elsie" id="link1">Elsie</a>,
<a class="sister" href="http://example.com/lacie" id="link2">Lacie</a>,
<a class="sister" href="http://example.com/tillie" id="link3">Tillie</a>]
```

.select()方法返回的是一个列表。

（2）通过 CSS 类名查找

```
print(soup.select(".sister"))
print(soup.select("[class~=sister]"))
```

输出结果：

```
[<a class="sister" href="http://example.com/elsie" id="link1">Elsie</a>, <a class="sister" href="http://example.com/lacie" id="link2">Lacie</a>, <a class="sister" href="http://example.com/tillie" id="link3">Tillie</a>]
[<a class="sister" href="http://example.com/elsie" id="link1">Elsie</a>, <a class="sister" href="http://example.com/lacie" id="link2">Lacie</a>, <a class="sister" href="http://example.com/tillie" id="link3">Tillie</a>]
```

（3）通过 ID 查找

```
print(soup.select("#link1"))
print(soup.select("a#link2"))
```

输出结果：

```
[<a class="sister" href="http://example.com/elsie" id="link1">Elsie</a>]
[<a class="sister" href="http://example.com/lacie" id="link2">Lacie</a>]
```

（4）通过是否存在某个属性来查找

```
print(soup.select('a[href]'))
```

输出结果：

```
[<a class="sister" href="http://example.com/elsie" id="link1">Elsie</a>, <a class="sister" href="http://example.com/lacie" id="link2">Lacie</a>, <a class="sister" href="http://example.com/tillie" id="link3">Tillie</a>]
```

（5）通过属性值来查找

```
print(soup.select('a[href="http://example.com/elsie"]'))
print(soup.select('a[href^="http://example.com/"]'))
print(soup.select('a[href$="tillie"]'))
print(soup.select('a[href*=".com/el"]'))
```

输出结果：

```
[<a class="sister" href="http://example.com/elsie" id="link1">Elsie</a>]
[<a class="sister" href="http://example.com/elsie" id="link1">Elsie</a>, <a class="sister" href="http://example.com/lacie" id="link2">Lacie</a>, <a class="sister"  href="http://example.com/tillie" id="link3">Tillie</a>]
[<a class="sister" href="http://example.com/tillie" id="link3">Tillie</a>]
[<a class="sister" href="http://example.com/elsie" id="link1">Elsie</a>]
```

2.9 爬虫常用类库 4：Selenium 操纵浏览器

Selenium 是一个自动化测试工具，支持各种浏览器，包括 Chrome、Safari、Firefox 等主流界面式浏览器。简单理解，Selenium 可以模拟操作浏览器，对一些需要动态加载的页面，不需要我们执行 JavaScript 等操作，即可自动加载完成后的页面。

2.9.1 安装 Selenium

直接使用 pip 安装：

```
pip install selenium
```

安装 Selenium 之后，还需要下载对应的浏览器驱动，放入系统路径中：

- Chrome driver：https://sites.google.com/a/chromium.org/chromedriver/home。
- Firefox driver：https://github.com/mozilla/geckodriver/releases。
- IE driver：https://github.com/mozilla/geckodriver/releases。
- Edge driver：https://developer.microsoft.com/en-us/microsoft-edge/tools/webdriver。

检查一下，打开 Python 控制台：

```
>>>from selenium import webdriver
>>>driver=webdriver.Chrome()
>>>driver.get('http://www.bing.com')
```

若能正常打开浏览器，则设置正确。

2.9.2 Selenium 的基本使用方法

先来看一段代码：

```
01 from selenium import webdriver
02 import time
03
04 driver = webdriver.Chrome()
05 driver.maximize_window()
06 driver.get(' https://cn.bing.com/')
07 driver.find_element_by_css_selector("#sb_form_q").
                                              send_keys("selenium")
08 driver.find_element_by_css_selector("#sb_form_go").click()
09 time.sleep(2)
10 driver.quit()
```

运行一下代码，查看效果。可以看到，浏览器先打开了 https://cn.bing.com/，然后窗口最大化，之后在搜索框输入了 selenium，然后进行搜索，显示搜索结果，两秒之后，浏览器关闭，如图 2.17 和图 2.18 所示。

再来分析代码：第 01 行先从 selenium 库导入 webdriver 模块，浏览器的基本操作都在这个模块中。第 04 行获取一个 Chrome 浏览器驱动，代表之后使用 Chrome 进行操作。第 05 行.maximize_window()的作用是浏览器窗口最大化。第 06 行.get()方法接收一个 URL 字符串，之后浏览器打开 URL。第 07 行使用了一个定位元素的方法.find_element_by_css_selector()来定位到输入框，并通过.send_keys()输入查询数据“selenium”。第 08 行定位到查询按钮，使用 click()方法实现单击操作，显示搜索结果。第 09 行等待两秒之后，第 10 行使用.quit()方法关闭浏览器。

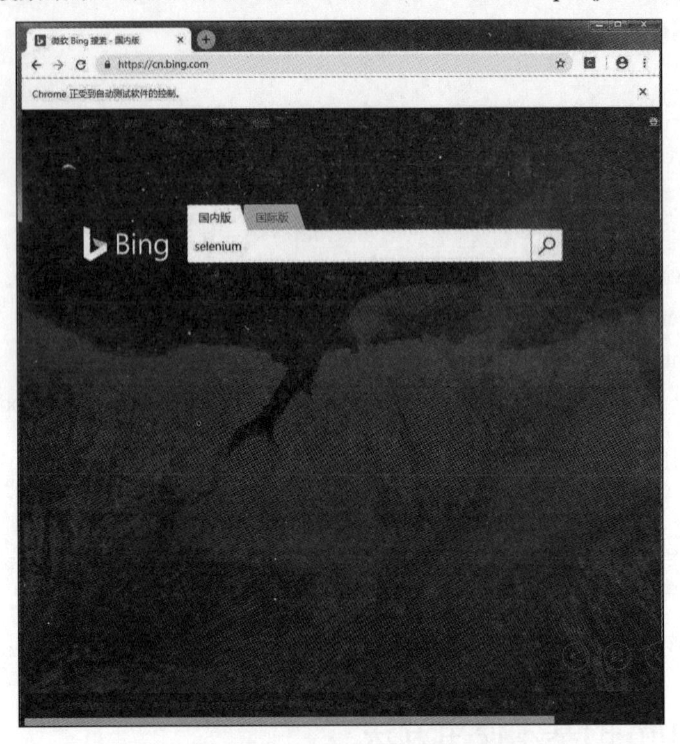

图 2.17　Selenium 打开浏览器

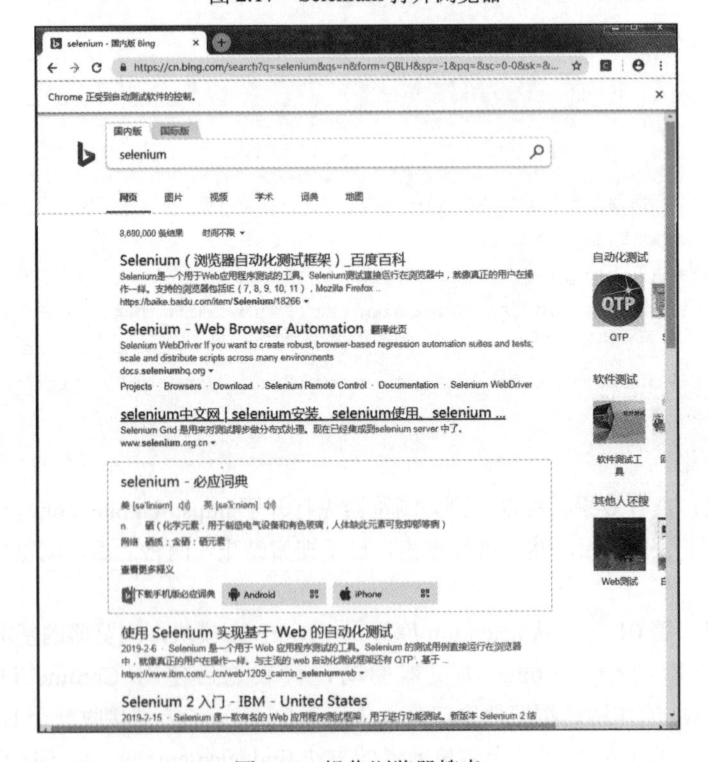

图 2.18　操作浏览器搜索

2.9.3　Selenium Webdriver 的原理

通过前面的例子，本小节简单介绍 Selenium 的运行原理。我们将 2.9.2 小节的代码运行分成 3 部分：

- 浏览器
- driver
- client

client 就是我们写的代码，我们无须知道浏览器具体的运行原理，只需要调用 driver，而 driver 知道如何驱动浏览器运行。在 Selenium 启动以后，driver 其实充当了 HTTP Server 服务器的角色，负责 client 和浏览器通信。client 根据 Webdriver 协议发送请求给 driver，driver 解析请求后，在浏览器上执行相应的操作，并把执行结果返回给 client。其中 Webdriver 协议包含几乎所有与浏览器的交互操作。通过这些协议，client 就可以通知 driver 执行哪些操作。

2.9.4　Selenium 中的元素定位方法

Selenium 支持 8 种元素定位方法：

- find_element_by_id()
- find_element_by_name()
- find_element_by_class_name()
- find_element_by_tag_name()
- find_element_by_link_text()
- find_element_by_partial_link_text()
- find_element_by_xpath()
- find_element_by_css_selector()

下面通过必应主页（http://cn.bing.com）来演示。我们定位到搜索输入框，如图 2.19 所示。

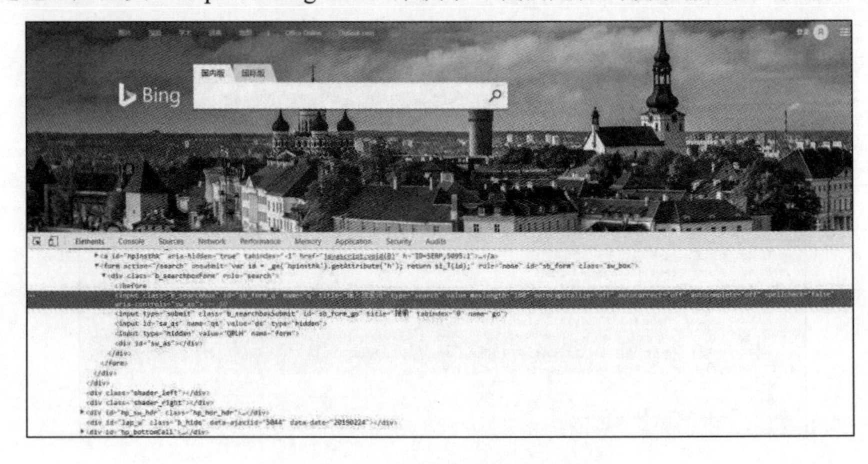

图 2.19　必应主页输入框定位

输入框的标签内容为：

```
<input class="b_searchbox" id="sb_form_q" name="q" title="输入搜索词" type="search" value="" maxlength="100" autocapitalize="off" autocorrect="off" autocomplete="off" spellcheck="false" aria-controls="sw_as">
```

下面使用不同的方法来定位。

通过 id 定位输入框：

```
driver.find_element_by_id("sb_form_q")
```

通过 name 定位输入框：

```
driver.find_element_by_name("q")
```

通过 classname 定位输入框：

```
driver.find_element_by_class_name("b_searchbox")
```

通过 tagname 定位输入框：

```
driver.find_element_by_tag_name("input")
```

通过 xpath 定位输入框：

```
driver.find_element_by_xpath("//input[@id='sb_form_q']")
```

通过 css 选择器定位输入框：

```
driver.find_element_by_css_selector("#sb_form_q")
```

对于文本链接的定位，有两种方法。本例以需要定位的文本“Outlook.com”标签来演示，其在浏览器开发工具中的内容如图 2.20 所示。

```
<a aria-owns="off_menu_cont" aria-controls="off_menu_cont" aria-expanded="false" target="_blank" onclick="hpulc4hdr();" href="https://outlook.com/?WT.mc_id=O16_BingHP?mkt=zh-CN" h="ID=SERP,5018.1">Outlook.com</a>
```

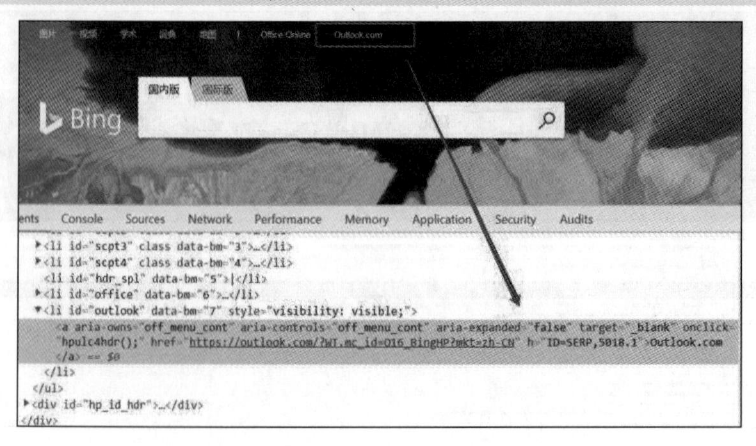

图 2.20　必应主页 Outlok.com 链接文本

（1）通过完整链接文本 link_text 定位文本链接：

```
driver.find_element_by_link_text("Outlookcom")
```

（2）通过 partial_linik_text 定位文本链接：

```
driver.find_element_by_ partial_linik_text("Outlook")
```

2.9.5 Selenium Webdriver 基本操作

定位到元素之后，需要进行一些相应的操作，Webdriver 支持的操作有很多，这里只介绍常用的一些方法，如表 2-24 所示。

表2-24　Webdriver常用方法

方法	说明	示例
（1）浏览器操作		
get()	访问 URL	driver.get("http://bing.com")
back()	后退上一步	driver.get(url1) driver.get(url2) driver.get(url3) driver.back()返回到 url2
forward()	前进下一步	driver.get(url1) driver.get(url2) driver.get(url3) driver.back()返回到 url2 driver.forward()返回到 url3
quit()	退出驱动，关闭所有窗口	driver.quit() 当打开多个窗口时，会关闭所有的窗口
close()	关闭当前窗口	driver.close() 关闭当前打开的窗口
maximize_window()	浏览器最大化	driver. maximize_window()
refresh	刷新浏览器	driver.refresh() 刷新浏览器
（2）元素操作		
send_keys()	向文本框类型输入数据	driver.find_element_by_tag('input').send_keys('selenium') 向 input 输入框中输入'selenium'
clear()	清空输入的数据	driver.find_element_by_tag('input').clear() 清空 input 输入框中的内容
click()	单击事件	driver.find_element_by_tag('button').click() 单击 button 按钮

（续表）

方法	说明	示例
enter()	触发键盘 enter 操作	driver.find_element_by_tag('input').send_keys('selenium') driver.find_element_by_tag('input').enter() 在 input 输入框中输入'selenium'，然后进行回车操作
text()	获取元素的文本内容	driver.find_element_by_ partial_link_text("Outlook").text() 获取 Outlook.com 文本
page_source	获取页面 HTML 内容	driver.get('http://www.bing.com') driver.page_source() 获取 Bing 首页页面 HTML
（3）Cookie 操作		
get_cookies	获取当前页的所有 Cookies	driver.get('http://www.bing.com') driver.get_cookies() 获取 Bing 首页的所有 Cookies
get_cookie(name)	获取当前页 Cookies 中的指定 name 值的 Cookie	driver.get_cookie('time') 获取 cookies 中 name=time 的 Cookie
add_cookie	添加 Cookie	driver.add_cookie({'time': '20180919123456'}) 添加一条 name=time，value=20180919123456 的 Cookie
delete_cookie(name)	删除一条 Cookie	driver.delete_cookie('time') 删除 name=time 的 Cookie
delete_all_cookies	删除所有 Cookie	driver.delete_cookies() 删除所有已获得的 Cookie

2.9.6 Selenium 实战：抓取拉钩网招聘信息

下面通过一个 Selenium 例子来演示使用方法。

【示例 2-13】抓取拉勾网招聘信息

在拉勾网输入关键字，查找招聘信息，主要抓取的数据如图 2.21 所示。

lagou.py 代码如下：

```
01 from selenium import webdriver
02 from bs4 import BeautifulSoup
03 import time
04
05
06 class Lagou:
07     def __init__(self):
08         # 定义浏览器驱动
09         self.driver = webdriver.Chrome()
10         self.driver.maximize_window()
```

```
11          # 起始网址
12          self.url = "https://www.lagou.com/"
13
14      def search(self, keywords):
15          # 打开页面
16          self.driver.get(self.url)
17          # 关闭弹窗
18          self.driver.find_element_by_xpath("//a[@class='tab focus']").
                                                          click()
19
20          # 在搜索框中输入关键字
21          self.driver.find_element_by_css_selector("#search_input").
                                                          send_keys(keywords)
22          # 单击按钮
23          self.driver.find_element_by_xpath ("//input[@id=
                                                          'search_button']").click()
24          # 等待两秒
25          time.sleep(2)
26          # 获取页面 html
27          page_source = self.driver.page_source
28          # 关闭浏览器
29          self.driver.quit()
30
31          return page_source
32
33      def get_jobs(self, page_source):
34          # 使用 BeautifulSoup 解析页面
35          soup = BeautifulSoup(page_source, 'lxml')
36          # 获取所有的招聘条目
37          hot_items = soup.select('.con_list_item')
38          for item in hot_items:
39              d = dict()
40              # 获取工作岗位名称
41              d['job'] = item.select_one(".position_link > h3").get_text()
42              # 获取公司名称
43              d['company'] = item.select_one(".company_name > a").get_text()
44              # 获取薪资
45              d['salary'] = item.select_one(".money").get_text()
46              print(d)
47
48
49  if __name__ == "__main__":
50      hot = Lagou()
```

```
51      # 搜索关键字
52      page_source = hot.search('python')
53      hot.get_jobs(page_source)
54
```

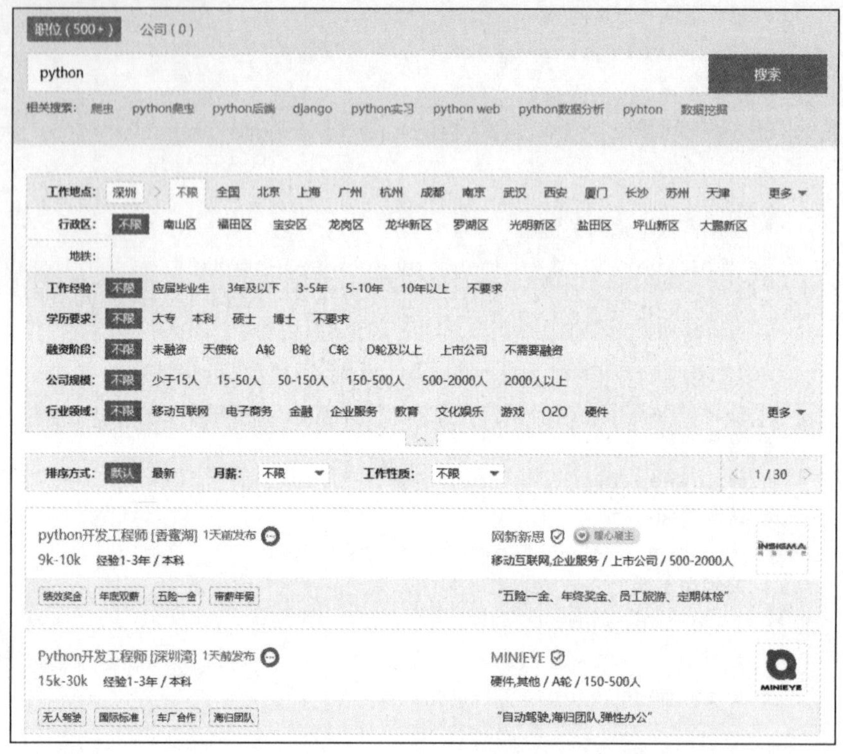

图 2.21　拉勾网招聘信息

在进入拉勾网时，会有一个选择地区的弹窗，首先需要单击默认地区以关闭弹窗（第 18 行）。然后定位到输入框，输入关键字“python”（第 21 行），然后单击按钮进行搜索（第 23 行）。之后获取到页面的 html 数据，在 get_job()方法中使用 BeautifulSoup 进行数据解析。提取到工作岗位、公司、薪资数据并进行输出。

运行结果：

```
{'job': 'python 开发工程师', 'company': '网新新思', 'salary': '9k-10k'}
{'job': 'Python 开发工程师', 'company': 'MINIEYE', 'salary': '15k-30k'}
{'job': 'Python 后台开发经理', 'company': 'AfterShip', 'salary': '30k-50k'}
{'job': 'python', 'company': 'Apier', 'salary': '10k-20k'}
{'job': 'python 开发工程师', 'company': '房讯通', 'salary': '15k-20k'}
{'job': 'Python 开发工程师', 'company': '联易融 linklogis', 'salary': '13k-25k'}
{'job': 'python 开发', 'company': '海万科技', 'salary': '10k-18k'}
{'job': 'Python 后台开发工程师', 'company': '华为', 'salary': '15k-25k'}
{'job': 'pyhton 开发', 'company': '比茨信息', 'salary': '7k-9k'}
{'job': 'Python 开发主管', 'company': '晨星 Morningstar', 'salary': '15k-20k'}
{'job': 'Python 工程师（运维平台）', 'company': 'OPPO', 'salary': '15k-30k'}
{'job': 'python 开发工程师', 'company': '深信服科技集团', 'salary': '11k-22k'}
```

```
{'job': 'Python开发工程师（安全）', 'company': '极光', 'salary': '15k-20k'}
{'job': 'python开发工程师', 'company': '盛迪嘉支付', 'salary': '15k-20k'}
```

2.10　爬虫常用类库 5：Scrapy 爬虫框架

Scrapy 是一个使用 Python 实现的，为了爬取网站数据、提取结构性数据而编写的应用框架，用途非常广泛。只需要定制开发几个模块就可以轻松地实现一个爬虫，用来抓取网页内容以及各种图片，非常方便。Scrapy 使用了 Twisted 异步网络框架来处理网络通信，可以加快下载速度，并且包含各种中间件接口，可以灵活地完成各种需求。

2.10.1　安装 Scrapy

Scrapy 基于 Python 语言，理所当然可以跨平台使用，不过相比于 Linux，Scrapy 在 Windows 上安装较为复杂。

1. 在 Windows 上安装 Scrapy

Scrapy 在 Windows 上的安装步骤比较复杂，由于 Scrapy 依赖包比较多，而有的包会有非 Python 依赖环境，因此 Windows 平台下直接使用 pip 安装大概率会报错。需要遵循以下步骤安装：

（1）在 https://www.lfd.uci.edu/~gohlke/pythonlibs/下载对应 Python 版本与系统的 Twisted 包，使用 pip 进行安装。

（2）在 https://www.lfd.uci.edu/~gohlke/pythonlibs/下载对应 Python 版本与系统的 lxml 包，使用 pip 进行安装。

（3）使用 pip install scrapy 进行安装，安装完成后，在命令行输入 scrapy，检测安装是否成功，若未报错，则安装成功：

```
C:\>scrapy
Scrapy 1.5.0 - no active project

Usage:
  scrapy <command> [options] [args]

Available commands:
  bench         Run quick benchmark test
  fetch         Fetch a URL using the Scrapy downloader
  genspider     Generate new spider using pre-defined templates
  runspider     Run a self-contained spider (without creating a project)
  settings      Get settings values
  shell         Interactive scraping console
  startproject  Create new project
  version       Print Scrapy version
```

```
  view         Open URL in browser, as seen by Scrapy

  [ more ]      More commands available when run from project directory

Use "scrapy <command> -h" to see more info about a command
```

2. 在 Ubuntu 上安装 Scrapy

在 Ubuntu 上安装 Scrapy 非常简单，直接使用 pip3 install scrapy 进行安装即可。

2.10.2 Scrapy 简介

Scrapy 包含爬虫所有的单元，并且提供了很多有用的中间件，可自由定制，方便使用。Scrapy 框架如图 2.22 所示。

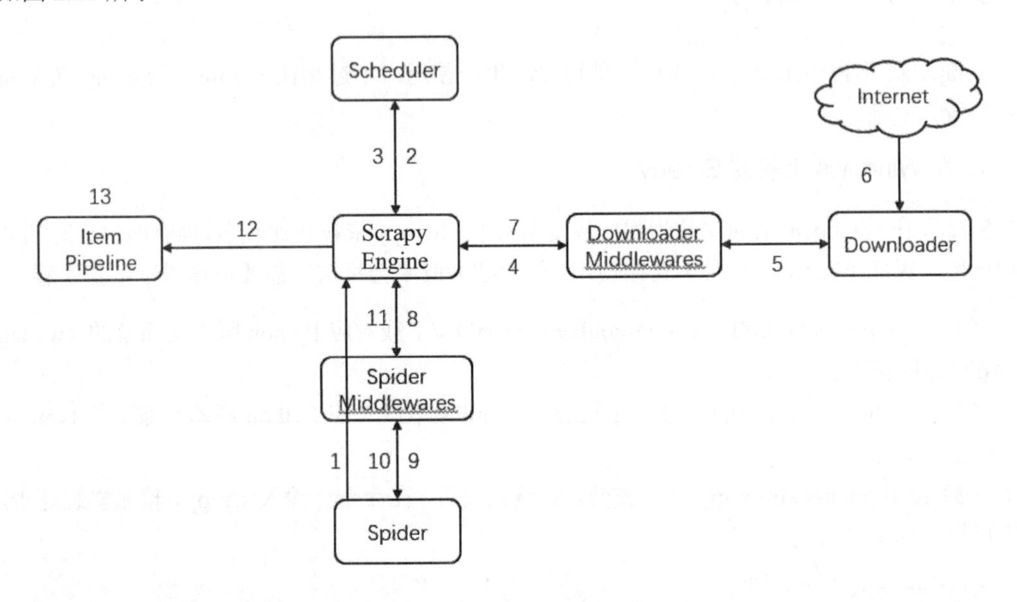

图 2.22　Scrapy 框架

下面对框架组件进行介绍。

- Scrapy Engine（引擎）：负责所有组件的数据传递，触发相应操作。
- Scheduler（调度器）：负责接收引擎发送的 Request 请求并排入队列，当引擎请求 Request 时再传递给引擎。
- Downloader（下载器）：接收引擎传递的 Request 请求并下载页面数据，然后将其获取到的 Response 传递给引擎，由引擎传递给 Spider 进行处理。
- Spider（爬虫）：用户主要编写的爬虫文件就是此部分文件，它负责处理引擎传递过来的 Response，从中分析提取数据，获取的 Item 字段传递给 Item Pipeline 处理，需跟进的 URL 传递给引擎，经由引擎传递给 Scheduler（调度器）。
- Item Pipeline（管道）：负责处理从 Spider 中获取到的 Item，进行过滤、存储等操作。

- Downloader Middlewares（下载中间件）：主要处理Scrapy引擎与下载器之间的请求及响应。
- Spider Middlewares（Spider中间件）：主要工作是处理蜘蛛的响应输入和请求输出。

了解基本架构及组件功能后，现在说明工作流程。

（1）Scrapy引擎从Spider获取起始Request。

（2）Scrapy引擎将获取到的Request发给调度中心排队入列。

（3）Scrapy引擎从调度中心请求获取需要处理的Request。

（4）Scrapy引擎获取到需处理的Request后，将Request发给下载器。

（5）Request在传递给下载器的过程中会经过下载器中间件，对Request进行处理。

（6）下载器根据Request从Internet下载内容，封装成Response对象传递给Scrapy引擎。

（7）下载器将Response传递给Scrapy引擎时，也会经过下载器中间件，对Response进行处理。

（8）Scrapy引擎将接收到的Response传递给Spider进行处理。

（9）Response传递给Spider的过程中，会经过Spider中间件，对Response进行处理。

（10）Spider接收Response，处理完之后会生成一个包含需要继续爬取的网址的Request和一个Item对象组成的Result，将Result传递给Scrapy引擎。

（11）Result传递给Scrapy的过程中会再次经过Spider中间件进行相应的处理。

（12）Scrapy引擎获取到Spider传递的Result，将其中的Item发送给Item Pipeline处理，将其中的Request发给调度器排队入列。

（13）Item Pipeline会对数据进行进一步的处理，包括整理、保存等。

重复这些步骤，就形成了一个完整的数据抓取处理流程。可以看到，Scrapy框架与前面介绍的一般爬虫一致，也包含调度器、下载器等对应的组件。

2.11 基本爬虫实战：抓取cnBeta网站科技类文章

前面我们分析了爬虫的基本原理，了解了一般的爬虫框架与封装好的Scrapy框架的运行流程之后，下面通过一个例子来加深对Scrapy爬虫框架的理解。

【示例2-14】使用通用爬虫框架来抓取cnBeta网站科技类文章

主要抓取的内容有文章标题、文章链接、文章发表日期这些数据，将抓取到的数据保存到本地文件中，如图2.23所示。

2.1节我们分析过爬虫的基本过程，知道爬虫框架主要的组成部分如下。

- URL管理器：负责管理待爬取的网页URL。
- 数据下载器：根据URL下载数据。
- 数据分析器：分析筛选下载的数据。
- 数据保存器：将筛选出的数据保存到文件或数据库。
- 调度器：负责整个系统的调度。

图 2.23 cnBeta 文章

下面开始编写这 5 个核心部分。

2.11.1 URL 管理器

URL 管理器的主要功能是收集、管理 URL 信息，包括待爬取的 URL、已经爬取过的 URL。程序 urlmanager.py 完整代码如下：

```
01 class URLmanager(object):
02     def __init__(self):
03         # 初始化待爬取 URL 与已爬取 URL 集合
04         self.new_urls = set()
05         self.old_urls = set()
06
07     def save_new_url(self, url):
08         # 将单条 URL 保存到待爬取集合中
09         if url is not None:
10             if url not in self.new_urls and url not in self.old_urls:
11                 print("保存新 URL:{}".format(url))
12                 self.new_urls.add(url)
13
14     def save_new_urls(self, url_list):
15         # 批量保存 URL
16         for url in url_list:
17             self.save_new_url(url)
18
19     def get_new_url(self):
```

```
20          # 取出一条未爬取的 URL，同时保存到已爬取的 URL 中
21          if self.get_new_url_num() > 0:
22              url = self.new_urls.pop()
23              self.old_urls.add(url)
24              return url
25          else:
26              return None
27
28      def get_new_url_num(self):
29          # 返回未爬取的 URL 数量
30          return len(self.new_urls)
31
32      def get_old_url_num(self):
33          # 返回已经爬取的 URL 数量
34          return len(self.old_urls)
```

【代码分析】

（1）首先初始化两个 set()数据类型：待爬取 new_urls 与已爬取 old_urls。使用 set 的好处是，set 存储数据唯一，不会保存重复的 URL，从而避免重复抓取数据。

（2）在保存新的 URL 时，需要判断是否包含在待爬取集合中，如果不存在，再判断是否包含在已爬取集合中，两个集合中都不包含此 URL，才将其存储至 new_urls 中。同时加了一条打印语句，以便在输出中查看具体 URL 信息，更好的方法是使用日志进行处理，后续章节会有介绍。

（3）从 new_urls 中取出一条 URL 进行爬取时，同时要将此 URL 保存至 old_urls。

（4）我们还定义了两个方法：get_new_url_num()与 get_old_url_num()，分别返回 new_urls 与 old_urls 中 URL 的数量。

2.11.2　数据下载器

数据下载器的作用是根据提供的 URL 下载网页数据，这里使用 requests 包进行下载操作。

程序 htmldownloader.py 代码如下：

```
01 import requests
02
03 class HtmlDownloader:
04
05     def download(self,url):
06     # 判断 URL 是否为空
07     if url is None:
08         return None
09         print("开始下载数据，网址{0}".format(url))
10         response = requests.get(url)
11         # 如果请求成功，则返回网页数据，否则返回 None
12         if response.status_code == 200:
```

```
13            print("下载数据成功")
14            # 指定使用 UTF-8 编码
15            response.encoding = 'utf-8'
16            return response.text
17        return None
18
19
20 if __name__ == "__main__"003A
21     url = 'http://www.bing.com/'
22     d = HtmlDownloader()
23     bing_html = d.download(url)
24     print(bing_html)
```

下载器功能很简单，只需使用 requests 中的 get()进行网页下载，在数据返回时则需要判断是否下载成功，即判断响应码是否为 200。

2.11.3 数据分析器

数据分析器接收下载到的网页数据，从中提取到我们需要的数据。这里使用 BeautifulSoup 进行页面解析、数据提取，主要提取内容为文章标题、文章链接、文章发布日期。

打开文章页面之后，使用浏览器开发者工具查看页面元素，如图 2.24 所示。

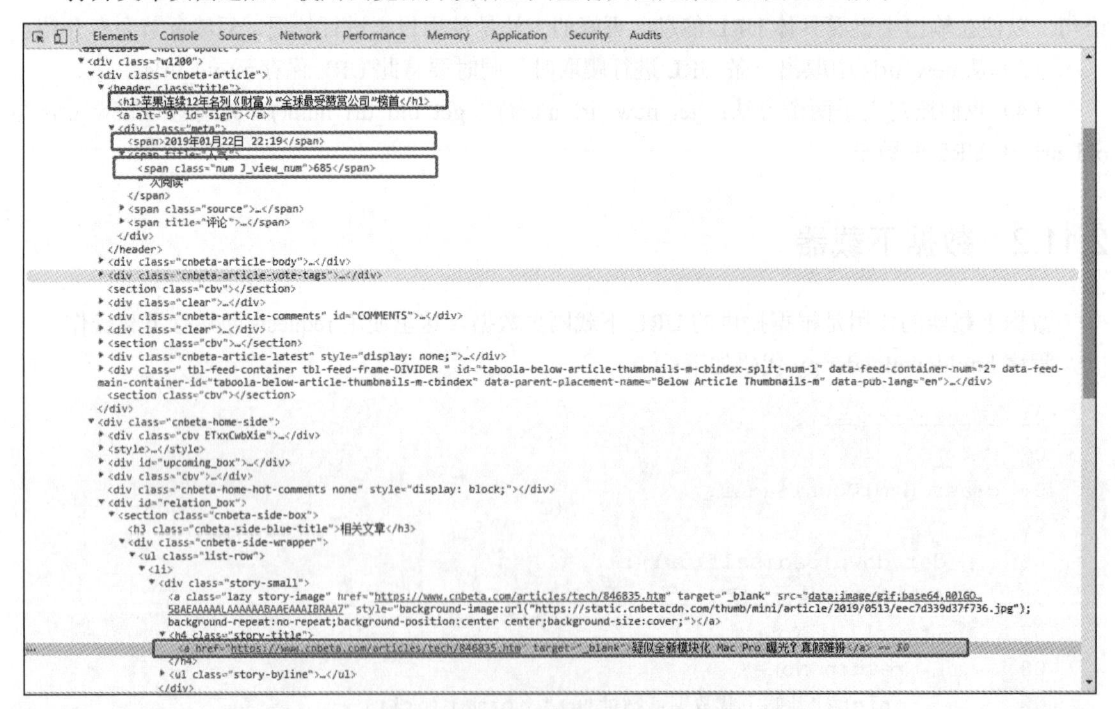

图 2.24 查看元素标签

查看页面元素可以看到，标题位于<div class="cnbeta-article"><header class="title"><h1>下，发布日期位于<div class="cnbeta-article"><header class="title"><div class="meta">中第一个<span>标签

下。对于文章 URL 的提取，通过检查页面中相关文章新闻模块中的文章标题元素信息，可以看到页面中科技类文章的标签为<a href=https://www.cnbeta.com/articles/tech/846835.htm target="_blank">疑似全新模块化 MacPro 曝光？真假难辨</a>形式，所以只需选择所有 href 属性中包含“tech”字段的 a 标签即可，再提取 href 属性字段值。

数据分析器 htmlparse.py 程序代码如下：

```
01 from bs4 import BeautifulSoup
02 from datasave import DataSave
03 import re
04 import requests
05
06 class HtmlParse:
07     # 主输出方法，返回提取的 URL 列表与待保存的数据
08     def parse_data(self,page_url,data):
09         print("开始分析提取数据")
10         # 如果待分析的文章的 URL 或者数据为空，则不做处理
11         if page_url is None or data is None:
12             return
13         soup =  BeautifulSoup(data,'lxml')
14         # 分别调用 get_urls()与 get_data()获取数据
15         urls = self.get_urls(soup)
16         data = self.get_data(page_url,soup)
17         return urls,data
18
19     # 提取科技类文章 URL
20     def get_urls(self,soup):
21         urls = list()
22         # 获取科技类文章地址 Tag
23         links = soup.select('a[href*="/tech/"]')
24         for link in links:
25             # 从 Tag 中提取网址数据
26             url = link['href']
27             urls.append(url)
28         return urls
29
30     # 提取文章数据
31     def get_data(self,page_url,soup):
32         data = {}
33         # 将文章的地址、标题、发布日期保存到字典中
34         # 文章 URL 只是使用参数 url
35         data['url'] = page_url
36         # select_one 获取符合条件的第一条
37         # 获取文章标题
38         title = soup.select_one('.cnbeta-article > header > h1')
39         # 获取发布日期
```

```
40          release_date = soup.select_one('.cnbeta-article > header >
                          .meta > span')
41          # 将数据保存到一个字典变量中
42          data['title'] = title.get_text()
43          data['release_date'] = release_date.get_text()
44          print("文章 url: {0}".format(page_url))
45          print("数据: {0}".format(data))
46          return data
47
48
49  if __name__ == "__main__":
50      url = 'https://www.cnbeta.com/articles/tech/811395.htm'
51      save = DataSave('D:\\Scrapy\\cnbeta.txt')
52      response = requests.get(url)
53      response.encoding = 'utf-8'
54      parse = HtmlParse()
55      u,d = parse.parse_data(url,response.text)
56      save.save(d)
57      print(u,d)
```

【代码分析】

（1）代码中定义两个方法：get_urls()与 get_data()，get_urls()返回一个 list 类型结果 urls，包含提取到的所有科技类文章 URL 数据；get_data()返回一个 dict 类型结果 data，包含提取到的文章数据。这里使用了 BeautifulSoup 中的 select_one()方法，只获取符合条件的第 1 条数据。

（2）分析器提供了一个对外输出数据的方法 parse_data()，返回从 get_urls()与 get_data()中提取到的 urls 与 data。

2.11.4 数据保存器

数据保存器的主要功能是保存从数据分析器中提取的数据。在本例中，我们将抓取到的数据保存到本地文件中，后面的章节将介绍如何保存至数据库中。

程序 datasave.py 代码如下：

```
01  import os
02  class DataSave:
03      # 指定数据保存的文件路径
04      def __init__(self,path):
05          self.path = path
06
07      def save(self,data):
08          # 判断文件路径是否存在，若不存在，则抛出错误
09          if not os.path.exists(self.path):
10              raise FileExistsError("文件路径不存在")
11          # 将数据写入文件中，已追加形式写入文件
```

```
12        with open(self.path,'a') as fp:
13            print("开始写入数据")
14            # 加上\n 换行写入数据
15            fp.write(str(data) + '\n')
16        fp.close()
17
18 if __name__ == "__main__":
19    test_data = 'this is a test,\n save it'
20    save_path = 'C:\\Users\\gstar\\Desktop\\file.txt'
21    ds = DataSave(save_path)
22    ds.save(test_data)
```

【代码分析】

（1）为了能在任意位置保存数据，在初始化时，我们需要先指定一个文件路径，在保存数据的时候，先判断文件路径是否存在，如果不存在，就抛出异常。

（2）打开文件，指定打开模式为 a，即以追加模式打开文件，可以不断地进行数据写入，因为我们是每抓取一篇文章数据立即写入文件，如果以 w 模式打开，则会覆盖前一次保存的数据，注意写入的数据要转化成 str 类型。

（3）最后做了一下测试，每一个类都写一个测试，这是一个好的习惯，读者可以试着在前面的 urlmanager.py 类文件中写一个测试。

2.11.5　调度器

调度器的作用是将所有组件关联起来，起到管理数据流通的作用。调度器 scheduler.py 程序代码如下：

```
01 from datasave import DataSave
02 from htmldownloader import HtmlDownloader
01 from htmlparse import HtmlParse
02 from urlmanager import URLManager
03
04 class Scheduler:
05    def __init__(self,path,root_url,count):
06        # 初始化各个组件
07        self.url_manager = URLManager()
08        self.data_save = DataSave(path)
09        self.html_parser = HtmlParse()
10        self.downloader = HtmlDownloader()
11        self.root_url = root_url
12        self.count = count
13
14    def run_spider(self):
15        # 先添加一条 URL 到未爬取 URL 集合中
```

```
16          self.url_manager.save_new_url(self.root_url)
17          # 判断：如果未爬取 URL 集合中还有网址，并且还没有爬取到 50 篇文章，那么继续爬取
18
19          while self.url_manager.get_new_url_num() and
                    self.url_manager.get_old_url_num() 20 < self.count:
21              try:
22                  # 获取一条未爬取 URL
23                  url = self.url_manager.get_new_url()
24                  # 下载数据
25                  response = self.downloader.download(url)
26                  # 分析数据，返回 URL 与文章相关数据
27                  new_urls,data = self.html_parser.parse_data(url,response)
28                  # 将获取到的 URL 保存到未爬取 URL 集合中
29                  self.url_manager.save_new_urls(new_urls)
30                  # 保存数据到本地文件
31                  self.data_save.save(data)
32                  print("已经抓取了{0}篇文章".format(len
                                        (self.url_manager.old_urls)))
33              except Exception as e:
34                  print("本篇文章抓取停止,{0}".format(e))
35
36
37  if __name__ == "__main__":
38      root_url = "https://www.cnbeta.com/articles/tech/811395.htm"
39      save_url = "D:\\Scrapy\\chap2\\2-1\\cnbeta.txt"
40      Spider = Scheduler(save_url,root_url,20)
41      Spider.run_spider()
```

【代码分析】

（1）首先初始化各个组件，其中数据存储器 DataSave 需要指定数据保存路径，即初始化方法 __init__()的 path 参数。另外，我们还需要指定起始 URL 和需要抓取的文章数据，分别对应参数 root_url 与 count。

（2）爬虫运行方法为 run_spider()。首先要给 URL 管理器传入起始 URL，之后通过一个 while 循环判断是否有新的 URL 及已抓取的 URL 数量，来决定是否继续进行抓取。

（3）使用 try…except…结构来避免在抓取过程中出错而中断流程。

运行爬虫，结果如图 2.25 所示。

```
开始下载数据，网址http://www.cnbeta.com/articles/tech/803949.htm
下载数据成功
开始分析提取数据
文章url: http://www.cnbeta.com/articles/tech/803949.htm
数据：{'url': 'http://www.cnbeta.com/articles/tech/803949.htm', 'title': '后悔药来了：微信7.0降到旧版方法一览', 'release_date': '2019年01月01日 16:07'}
保存新URL:https://www.cnbeta.com/articles/tech/803769.htm
保存新URL:https://www.cnbeta.com/articles/tech/803533.htm
保存新URL:https://www.cnbeta.com/articles/tech/803371.htm
保存新URL:https://www.cnbeta.com/articles/tech/803183.htm
保存新URL:https://www.cnbeta.com/articles/tech/803113.htm
保存新URL:https://www.cnbeta.com/articles/tech/801673.htm
保存新URL:https://www.cnbeta.com/articles/tech/794019.htm
开始写入数据
已经抓取了18篇文章
开始下载数据，网址http://www.cnbeta.com/articles/tech/808491.htm
下载数据成功
开始分析提取数据
文章url: http://www.cnbeta.com/articles/tech/808491.htm
数据：{'url': 'http://www.cnbeta.com/articles/tech/808491.htm', 'title': 'Purism推出包含4K显示屏的Librem 15 Linux笔记本电脑', 'release_date': '2019年01月15日 00:25'}
保存新URL:http://www.cnbeta.com/articles/tech/808183.htm
保存新URL:http://www.cnbeta.com/articles/tech/807963.htm
保存新URL:https://www.cnbeta.com/articles/tech/807215.htm
保存新URL:https://www.cnbeta.com/articles/tech/807497.htm
开始写入数据
已经抓取了19篇文章
开始下载数据，网址https://www.cnbeta.com/articles/tech/812701.htm
下载数据成功
开始分析提取数据
文章url: https://www.cnbeta.com/articles/tech/812701.htm
数据：{'url': 'https://www.cnbeta.com/articles/tech/812701.htm', 'title': 'FB工程师光脚进餐厅引内部文化冲突 销售人员:不能忍', 'release_date': '2019年01月26日 10:25'}
保存新URL:http://www.cnbeta.com/articles/tech/812577.htm
保存新URL:http://www.cnbeta.com/articles/tech/812489.htm
保存新URL:http://www.cnbeta.com/articles/tech/812425.htm
保存新URL:http://www.cnbeta.com/articles/tech/812379.htm
保存新URL:http://www.cnbeta.com/articles/tech/812265.htm
开始写入数据
已经抓取了20篇文章

Process finished with exit code 0
```

图 2.25　运行结果

保存的数据如图 2.26 所示。

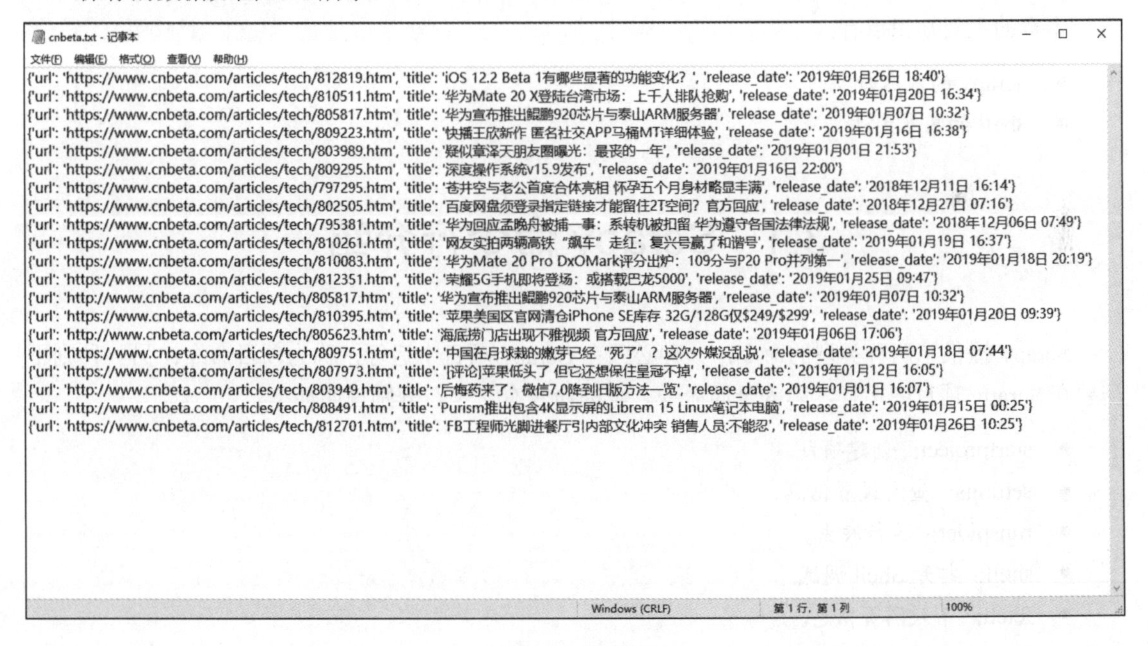

cnbeta.txt - 记事本

文件(F)　编辑(E)　格式(O)　查看(V)　帮助(H)

```
{'url': 'https://www.cnbeta.com/articles/tech/812819.htm', 'title': 'iOS 12.2 Beta 1有哪些显著的功能变化？', 'release_date': '2019年01月26日 18:40'}
{'url': 'https://www.cnbeta.com/articles/tech/810511.htm', 'title': '华为Mate 20 X登陆台湾市场：上千人排队抢购', 'release_date': '2019年01月20日 16:34'}
{'url': 'https://www.cnbeta.com/articles/tech/805817.htm', 'title': '华为宣布推出鲲鹏920芯片与泰山ARM服务器', 'release_date': '2019年01月07日 10:32'}
{'url': 'https://www.cnbeta.com/articles/tech/809223.htm', 'title': '快播王欣新作 匿名社交APP马桶MT详细体验', 'release_date': '2019年01月16日 16:38'}
{'url': 'https://www.cnbeta.com/articles/tech/803989.htm', 'title': '疑似章泽天朋友圈曝光：最丧的一年', 'release_date': '2019年01月01日 21:53'}
{'url': 'https://www.cnbeta.com/articles/tech/809295.htm', 'title': '深度操作系统v15.9发布', 'release_date': '2019年01月16日 22:00'}
{'url': 'https://www.cnbeta.com/articles/tech/797295.htm', 'title': '苍井空与老公首度合体亮相 怀孕五个月身材略显丰满', 'release_date': '2018年12月11日 16:14'}
{'url': 'https://www.cnbeta.com/articles/tech/802505.htm', 'title': '百度网盘须登录指定链接才能留住2T空间？官方回应', 'release_date': '2018年12月27日 07:16'}
{'url': 'https://www.cnbeta.com/articles/tech/795381.htm', 'title': '华为回应孟晚舟被捕一事：系转机被扣留 华为遵守各国法律法规', 'release_date': '2018年12月06日 07:49'}
{'url': 'https://www.cnbeta.com/articles/tech/810261.htm', 'title': '网友实拍两辆高铁“飙车”走红：复兴号赢了和谐号', 'release_date': '2019年01月19日 16:37'}
{'url': 'https://www.cnbeta.com/articles/tech/810083.htm', 'title': '华为Mate 20 Pro DxOMark评分出炉：109分与P20 Pro并列第一', 'release_date': '2019年01月18日 20:19'}
{'url': 'https://www.cnbeta.com/articles/tech/812351.htm', 'title': '荣耀5G手机即将登场：或搭载巴龙5000', 'release_date': '2019年01月25日 09:47'}
{'url': 'http://www.cnbeta.com/articles/tech/805817.htm', 'title': '华为宣布推出鲲鹏920芯片与泰山ARM服务器', 'release_date': '2019年01月07日 10:32'}
{'url': 'https://www.cnbeta.com/articles/tech/810395.htm', 'title': '苹果美国区官网清仓iPhone SE库存 32G/128G仅$249/$299', 'release_date': '2019年01月20日 09:39'}
{'url': 'http://www.cnbeta.com/articles/tech/805623.htm', 'title': '海底捞门店出现不雅视频 官方回应', 'release_date': '2019年01月06日 17:06'}
{'url': 'https://www.cnbeta.com/articles/tech/809751.htm', 'title': '中国在月球栽的嫩芽已经“死了”？这次外媒没乱说', 'release_date': '2019年01月18日 07:44'}
{'url': 'https://www.cnbeta.com/articles/tech/807973.htm', 'title': '[评论]苹果低头了 但它还想保住皇冠不掉', 'release_date': '2019年01月12日 16:05'}
{'url': 'http://www.cnbeta.com/articles/tech/803949.htm', 'title': '后悔药来了：微信7.0降到旧版方法一览', 'release_date': '2019年01月01日 16:07'}
{'url': 'http://www.cnbeta.com/articles/tech/808491.htm', 'title': 'Purism推出包含4K显示屏的Librem 15 Linux笔记本电脑', 'release_date': '2019年01月15日 00:25'}
{'url': 'https://www.cnbeta.com/articles/tech/812701.htm', 'title': 'FB工程师光脚进餐厅引内部文化冲突 销售人员:不能忍', 'release_date': '2019年01月26日 10:25'}
```

Windows (CRLF)　第 1 行，第 1 列　100%

图 2.26　保存结果

第 3 章 Scrapy 命令行与 Shell

第 2 章介绍了 Scrapy 的基本架构及原理，本章开始介绍 Scrapy 的基本使用方法以及常用的命令。Scrapy 是通过一个命令行工具 Shell 来控制相关动作行为的，通过 Shell 工具可以方便快捷地调试代码。因此，掌握 Scrapy 命令行与 Shell 可以使 Scrapy 的开发变得事半功倍。

本章的主要知识点有：

- Scrapy 的基本命令
- Shell 的基本用法

3.1 Scrapy 命令行介绍

Scrapy 提供了两种类型的命令：一种必须在 Scrapy 项目中运行，称为项目命令；另一种则不需要在 Scrapy 项目中运行，称为全局命令。全局命令有 7 个，分别说明如下。

- startproject：创建项目。
- settings：查看设置信息。
- runspider：运行爬虫。
- shell：打开 Shell 调试。
- fetch：下载网页信息。
- view：使用浏览器打开指定网址。
- version：查看版本号。

项目命令有 7 个，分别说明如下。

- crawl：运行指定的爬虫。
- check：检查爬虫代码。

- list：列出所有的爬虫。
- edit：使用默认的编辑器编辑爬虫文件。
- parse：使用爬虫抓取指定的 URL。
- genspider：创建爬虫。
- bench：快速的性能测试。

Scrapy 的命令涉及项目创建、爬虫创建、运行等具体事务。在命令行工具中输入 scrapy 将会列出这些命令，如图 3.1 所示。

```
命令提示符
D:\Scrapy>scrapy
Scrapy 1.5.0 - no active project

Usage:
  scrapy <command> [options] [args]

Available commands:
  bench         Run quick benchmark test
  fetch         Fetch a URL using the Scrapy downloader
  genspider     Generate new spider using pre-defined templates
  runspider     Run a self-contained spider (without creating a project)
  settings      Get settings values
  shell         Interactive scraping console
  startproject  Create new project
  version       Print Scrapy version
  view          Open URL in browser, as seen by Scrapy

  [ more ]      More commands available when run from project directory

Use "scrapy <command> -h" to see more info about a command

D:\Scrapy>
```

图 3.1　Scrapy 命令

每个命令都可以通过如图 3.2 所示的方式查看具体用法与参数。

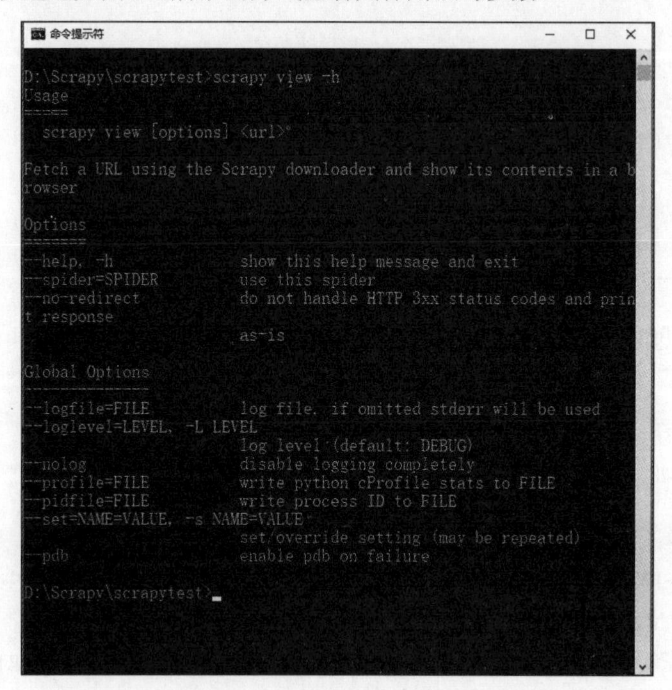

图 3.2　Scrapy 命令的帮助

本节将讲解常用命令的用法与功能。

3.1.1 使用 startproject 创建项目

最先要掌握的命令就是创建 Scrapy 项目的命令，格式为：

```
scrapy startproject <project_name> [project_dir]
```

这条命令将会在 project_dir 目录下创建一个项目，如果没有指定 project_dir，创建的目录就会与项目同名，效果如图 3.3 所示。

```
命令提示符

D:\Scrapy>scrapy startproject scrapytest
New Scrapy project 'scrapytest', using template directory 'D:\\SoftWare\\Ana
conda\\lib\\site-packages\\scrapy\\templates\\project', created in:
    D:\Scrapy\scrapytest

You can start your first spider with:
    cd scrapytest
    scrapy genspider example example.com

D:\Scrapy>
```

图 3.3 创建爬虫项目

创建项目之后，查看项目目录，如下所示：

```
──scrapytest
    │─scrapy.cfg
    └──scrapytest
        └──items.py
        │─middlewares.py
        │─pipelines.py
        │─settings.py
        │─__init__.py
        ├──spiders
           └──__init__.py
```

其中，scrapy.cfg 为全局配置文件，包含定义项目设置的 Python 模块，如下：

```
[settings]
default = myproject.settings
```

scrapytest 子文件夹中：

- item.py 中定义 Scrapy 的输出内容。
- middlewares.py 中定义各种中间件，主要为了处理各种 Request/Response。
- pipelines.py 中定义管道，功能为如何处理抓取到的数据。
- setting.py 为项目配置文件，所有的管道、中间件等其他参数必须在 setting.py 中激活才能生效。
- spiders 子文件夹中存放所有的爬虫文件。

3.1.2 使用 genspider 创建爬虫

创建项目框架之后，开始创建爬虫。该命令需要进入 scrapytest 目录执行，命令格式为：

```
scrapy genspider [-t template] <name> <domain>
```

其中-t 可以指定使用的爬虫模板，<name>为爬虫名称，<domain>用于生成 allowed_domains 和 start_urls Spider 属性值。

genspider 可以使用的模板有 4 个：

- basic
- crawl
- csvfeed
- xmlfeed

basic 为基本爬虫模板，crawl 模板生成继承 CrawlSpider 爬虫类的 Spider，csvfeed、xmlfeed 分别生成继承 CSVFeedSpider 与 XMLFeedSpider 爬虫类的 Spider，后面章节会讲到除 basic 以外的其余 3 种通用爬虫。本小节使用 crawl 模板创建爬虫。

【示例 3-1】Scrapy 创建爬虫

```
scrapy genspider -t crawl test test.com
```

genspider 命令如图 3.4 所示。

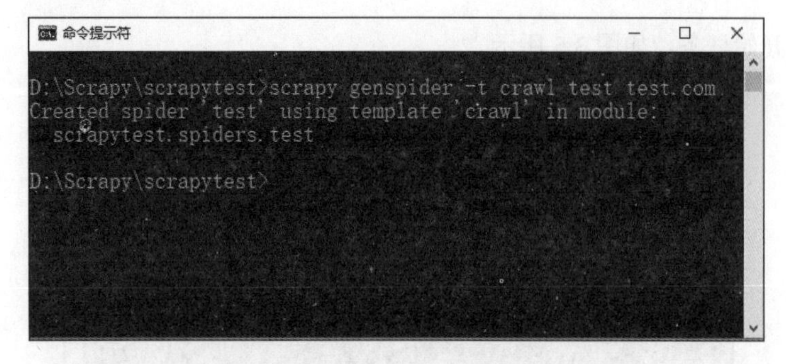

图 3.4　genspider 命令

此时会在 spider 目录下生成一个名为 test.py 的爬虫文件，内容为：

```
01 # -*- coding: utf-8 -*-
02 import scrapy
03 from scrapy.linkextractors import LinkExtractor
04 from scrapy.spiders import CrawlSpider, Rule
05
06
07 class TestSpider(CrawlSpider):
08     name = 'test'
09     allowed_domains = ['test.com']
```

```
10      start_urls = ['http://test.com/']
11
12      rules = (
13          Rule(LinkExtractor(allow=r'Items/'), callback='parse_item',
                                                         follow=True),
14      )
15
16      def parse_item(self, response):
17          i = {}
18          #i['domain_id'] = response.xpath
                                  ('//input[@id="sid"]/@value').extract()
19          #i['name'] = response.xpath('//div[@id="name"]').extract()
20          #i['description'] = response.xpath
                                  ('//div[@id="description"]').extract()
21          return i
```

可以看到，name 等属性已经生成，Rule（后面章节将会对 Rule 做详细介绍）也生成了一条，回调方法为 parse_item，可以根据需要进行修改。

3.1.3 使用 crawl 启动爬虫

启动爬虫的命令格式如下：

```
>>> scrapy crawl <spidername>
```

启动 Spider 开始任务，如图 3.5 所示。

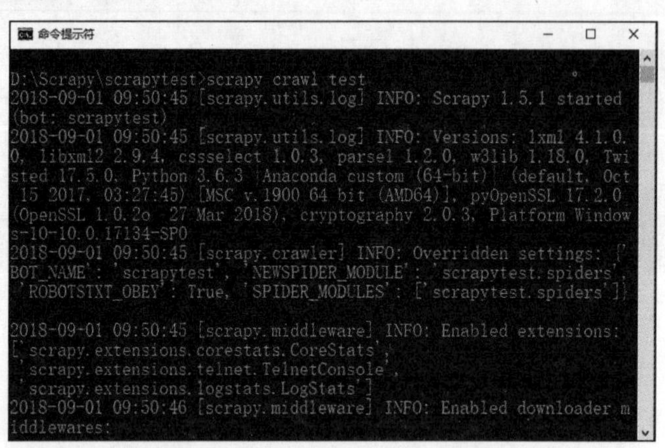

图 3.5 使用 crawl 命令启动爬虫

3.1.4 使用 list 查看爬虫

使用方法：

```
>>> scrapy list
```

list 命令很简单，列出项目的所有 Spider，如图 3.6 所示。

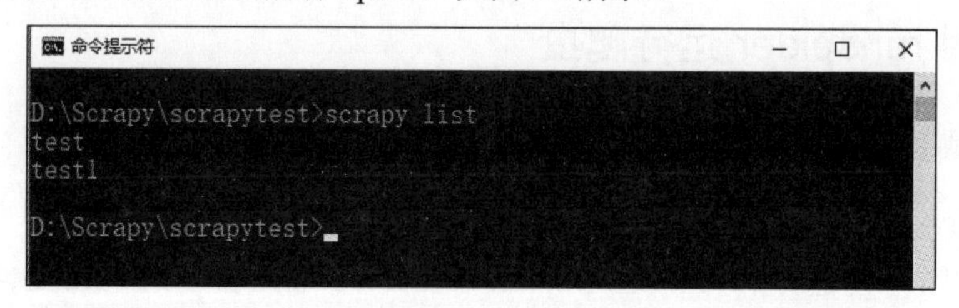

图 3.6　list 命令

3.1.5　使用 fetch 获取数据

fetch 使用方法：

```
>>> scrapy fetch [options] <url>
```

该命令使用 Scrapy 默认的下载器下载指定的 URL 页面。注意，如果在项目中运行此命令，会自动使用项目中 Spider 的相关设置，而在项目外运行则使用默认的设置。如图 3.7 所示，上半部分为在项目中运行命令，下半部分为在项目外运行命令。

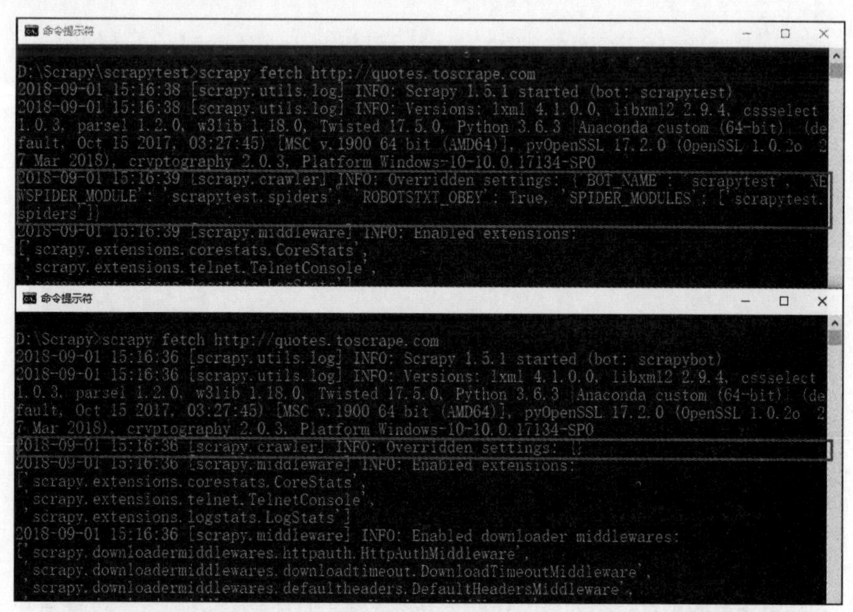

图 3.7　fetch 命令

另外，fetch 还有一些参数可以使用，分别说明如下。

- --spider=SPIDER：使用指定的 Spider 代替默认值。
- --headers：打印返回 Response 的 headers，默认打印 Response 的 body 部分。
- --no-redirect：取消重定向抓取，默认抓取重定向的 URL 数据。
- --nolog：取消日志的打印，直接显示数据。

3.1.6 使用 runspider 运行爬虫

runspider 使用方法：

```
>>> scrapy runspider [options] spider.py
```

该命令可以在未创建项目时直接运行 Spider 爬虫文件。它比较有用的参数是--output=FILE 或者-o FILE，将抓取结果保存到 FILE 文件中。

说明：runspider 命令适用于简单快速的爬虫任务。

【示例 3-2】runspider 运行爬虫文件 testspider.py

```
01 import scrapy
02 
03 class TestSpider(scrapy.Spider):
04     name = 'testspier'
05     start_urls = ['http://www.bing.com']
06 
07     def parse(self, response):
08         title = response.css('title::text').extract_first()
09         return({'title':title})
```

运行命令，结果如图 3.8 所示。

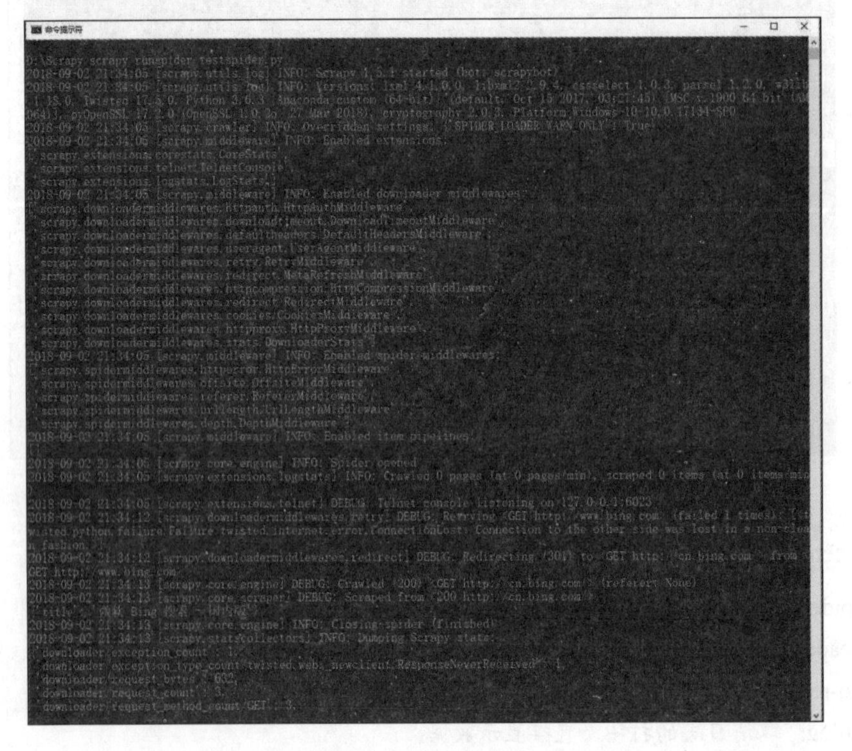

图 3.8 runspider 命令

3.1.7 通过view使用浏览器打开URL

view使用方法：

```
>>> scrapy view <url> [options]
```

该命令下载一个页面并使用默认浏览器打开。某些情况下，Scrapy 下载的页面与我们使用浏览器看到的页面并不一致，这时可以使用view命令查看对比。

3.1.8 使用parse测试爬虫

parse使用方法：

```
>>> scrapy parse [options] http://example.com/
```

parse是一个非常有用的命令，这个命令经常作为测试爬虫使用。可以为此命令指定Spider、Pipeline（管道）、回调函数等一系列爬虫参数，常用参数具体如下。

- --spider=SPIDER：使用特定的爬虫处理。
- --a NAME=VALUE：设定爬虫参数。
- --callback or -c：处理Response的回调函数。
- --meta or -m：为Request传的参数，例如 - meta='{"foo" : "bar"}'。
- --pipelines：使用管道处理Item。
- --rules or -r：使用CrawlSpider时指定的rules。
- --noitems：不显示爬取的Item。
- --nolinks：不显示解析的链接。

3.2 Scrapy Shell命令行

3.1节介绍的都是些基本命令。Scrapy中最常用的命令就是Shell命令了。在这里单独使用一节来介绍。

3.2.1 Scrapy Shell的用法

Scrapy Shell是一个交互终端，可以在未启动Spider的情况下调试代码。其本意是用来测试提取数据的代码，不过也可以将其作为正常的Python终端，在上面测试运行任何Python代码。

该终端用来测试XPath或CSS表达式，测试其能否从抓取的网页中正确地提取数据。因此，无须启动爬虫，我们可以边写代码边测试，无须每次修改代码之后测试整个爬虫来查看抓取结果。一旦掌握了Scrapy Shell，我们会发现开发和调试代码将无比简单。

Shell 的启动命令为：

```
>>> scrapy shell <url>
```

url 为待爬取的地址。当然，Shell 也可以打开本地文件，以下几种格式都是支持的：

```
>>> scrapy view <url> [options]
# UNIX-格式
scrapy shell ./path/to/file.html
scrapy shell ../other/path/to/file.html
scrapy shell /absolute/path/to/file.html
# 文件 URI
scrapy shell file:///D:/webfiles/bing.html
scrapy shell D:\webfiles\bing.html
```

读取本地网址的时候需要加上路径 ./，相对路径使用../。直接打开文件将会报错：scrapy shell index.html。

与普通 Python 控制台相比，Scrapy Shell 多了一些 Scrapy 爬虫中特有的功能，分别说明如下。

- shelp()：打印出所有可使用的属性与命令。
- fetch(url[, redirect=True])：从给定的 URL 获取一个新的 Response，同时更新所有相关的项目数据。当指定 redirect=False 时，不会获取重定向的数据。
- fetch(request)：根据给定的 Request 获取一个新的 Response，同时更新所有相关的项目数据。
- view(response)：使用指定的 Response 打开浏览器，方便检查抓取数据。当使用这条命令的时候，为使外部的图像或者表格等正确显示，会自动为 Response 中添加一个<base>标签，指定基准 URL，也就是 Response 对应的 URL。

使用 Scrapy Shell 下载页面时，会自动生成一些对象，分别说明如下。

- crawler：当前使用的 crawler。
- spider：处理当前页面使用的 spider，当没有指定 spider 时，则是一个 Spider 对象。
- request：获取最新页面所对应的 Request。
- response：获取最新页面所对应的 Response。
- settings：当前的 Scrapy 配置信息。

3.2.2 实战：解析名人名言网站

下面通过解析名人名言网站（http://quotes.toscrape.com/）来学习 Scrapy Shell 的用法，其中页面内容如图 3.9 所示。

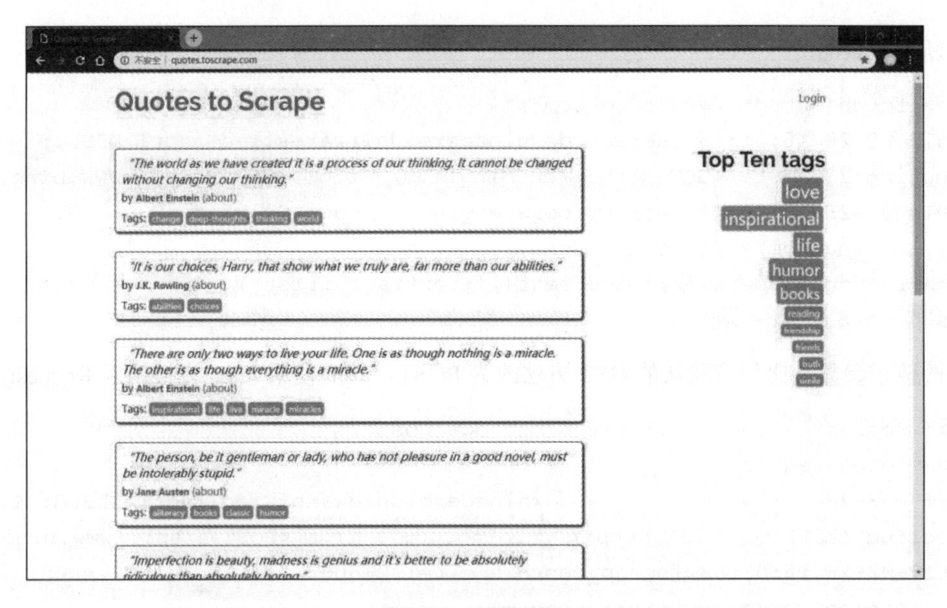

图 3.9 网页内容

（1）启动 Scrapy Shell，加上 --nolog 不打印日志，如图 3.10 所示。

```
命令提示符 - scrapy shell http://quotes.toscrape.com/ --nolog
C:\Users\test>scrapy shell http://quotes.toscrape.com/ --nolog
[s] Available Scrapy objects:
[s]   scrapy     scrapy module (contains scrapy.Request, scrapy.Selector, etc)
[s]   crawler    <scrapy.crawler.Crawler object at 0x0000027B5FEBAC18>
[s]   item       {}
[s]   request    <GET http://quotes.toscrape.com/>
[s]   response   <200 http://quotes.toscrape.com/>
[s]   settings   <scrapy.settings.Settings object at 0x0000027B62539BE0>
[s]   spider     <DefaultSpider 'default' at 0x27b627c37b8>
[s] Useful shortcuts:
[s]   fetch(url[, redirect=True]) Fetch URL and update local objects (by default, redirects
are followed)
[s]   fetch(req)                  Fetch a scrapy.Request and update local objects
[s]   shelp()           Shell help (print this help)
[s]   view(response)    View response in a browser
>>> _
```

图 3.10 –nolog 命令输出

可以看到，Scrapy Shell 使用 Scrapy Downloader 根据 URL 下载内容，并打印前面介绍过的可用对象与方法。

（2）下载页面之后，可以对下载内容进行检查、操作。

获取页面标题：

```
>>> response.xpath('//title/text()').extract_first()
'Quotes to Scrape'
>>> response.css('title::text').extract_first()
'Quotes to Scrape'
```

重新下载页面：

```
>>> fetch("http://www.bing.com")
2019-02-28 16:01:46 [scrapy.downloadermiddlewares.redirect] DEBUG:
Redirecting (301) to <GET http://cn.bing.com/> from <GET http://www.bing.com>
2019-02-28 16:01:47 [scrapy.core.engine] DEBUG: Crawled (200) <GET
http://cn.bing.com/> (referer: None)
>>> response.xpath('//title/text()').extract_first()
'微软 Bing 搜索 - 国内版'
```

修改请求参数，比如将默认的 GET 方法改为 POST，然后使用 fetch 直接请求 Request：

```
>>> request = request.replace(method='post')
>>> fetch(request)
2019-02-28 16:03:58 [scrapy.downloadermiddlewares.redirect] DEBUG:
Redirecting (301) to <POST http://cn.bing.com/> from <POST http://www.bing.com>
2019-02-28 16:03:58 [scrapy.core.engine] DEBUG: Crawled (200) <POST
http://cn.bing.com/> (referer: None)
打印 Response 中 headers 信息：
>>>from pprint import pprint
>>>pprint(response.headers)
{b'Cache-Control': [b'private, max-age=0'],
 b'Content-Type': [b'text/html; charset=utf-8'],
 b'Date': [b'Thu, 28 Feb 2019 08:03:59 GMT'],
 b'P3P': [b'CP="NON UNI COM NAV STA LOC CURa DEVa PSAa PSDa OUR IND"'],
 b'Vary': [b'Accept-Encoding'],
 b'X-Msedge-Ref': [b'Ref A: B16BBC6E7C214B0F918F44CE06DE0C97 Ref B:
BJ1EDGE03'
                  b'06 Ref C: 2019-02-28T08:04:00Z']}
```

第4章 Scrapy 爬虫

第 3 章介绍了 Scrapy 命令行基本命令与 Shell 模式的基本用法，本章将正式开始爬虫的讲解。首先介绍 Spider 的常用基本属性，再介绍 Spider 中常用的方法，紧接着对提取数据的选择器进行介绍，读者可参照示例进行理解。

本章知识点：

- Scrapy 基本类组件说明
- Scrapy 中的 Selector 选择器
- Scrapy 通用爬虫介绍与使用

4.1 编写爬虫

Scrapy 主要通过 Spider 类来实现爬虫的相关功能。通俗来讲，Spider 类定义了爬取某个或某些网站的规则，包括爬取数据与提取数据。

Spider 循环爬取步骤如下：

步骤01 通过 start_requests()以 start_urls 中的 URL 初始化 Request，下载完毕后返回 Response，作为参数传给回调函数 parse。

步骤02 使用 parse 函数分析 Response，可以返回 Item 对象、dict、Request 或一个包含三者的可迭代容器。其中，Request 可以经过 Scrapy 继续下载内容，调用设置的回调函数。

步骤03 在 parse 函数内，使用 Selector 分析 Response，提取相应数据。

4.1.1 scrapy.Spider 爬虫基本类

编写的爬虫 Spider 是通过继承 scrapy.Spider 类来实现的。Spider 类是最简单的爬虫类。每个其他的 Spider 必须继承自该类（包括 Scrapy 自带的其他 Spider 以及自定义编写的 Spider）。Spider 类没有提供什么特殊的功能，仅仅提供了 start_requests() 的默认实现，读取并请求 Spider 属性中的 start_urls，并根据返回的结果（Resulting Responses）调用 Spider 的 parse 方法。

Spider 类常用的属性如下。

- name

对一个 Spider 来说，name 属性是必须且唯一的，其定义了 Spider 的名字，而 Scrapy 通过 Spider 的名字来定位并且初始化 Spider。

- allowed_domains

该属性可选，其包含允许 Spider 爬取的域名列表。当中间件 OffsiteMiddleWare 启用时，将不会跟进不在列表中的域名。

- start_urls

start_urls 是一个 URL 列表。当没有指定特定 URL 时，Spider 将从该列表中开始获取页面数据，后续的 URL 将从获取的数据中提取。

- custom_settings

该属性可选，是一个 dict。当 Spider 启动时，会覆盖项目的设置，由于设置必须在初始化前被更新，因此必须设定为 class 属性。

4.1.2 start_requests()方法

先看 start_requests()常用的方法定义：

```
01 def start_requests(self):
02     urls = 
["http://www.test1.com/","http://www.test2.com/","http://www.test3.com/"]
03     for url in urls:
04         yield scrapy.Request(url=url, callback=self.parse)
```

在爬虫启动时，Scrapy 会调用 start_requests()方法，从 urls（第 02 行）表中依次获取 url，为此 url 生成 Request（第 04 行），然后调用回调方法 parse（第 04 行）处理生成的 Request。由于此方法只会调用一次，因此可以将此方法写成一个生成器。

start_requests()默认返回最普通的 Request，在使用时可以重写此方法以满足我们的需求，比如在爬虫启动时登录某个网站可以使用 FormRequest：

```
01 def start_requests(self):
02    return [scrapy.FormRequest("http://www.example.com/login",
03                          formdata={'user': 'john', 'pass': 'secret'},
04                          callback=self.login)]
```

FormRequest 的具体使用方法在后面章节会有详细介绍。

4.1.3 parse(response)方法

parse()方法基本写法：

```
01 def parse(self, response):
02     for media in response.css("li.media"):
03         yield{
04             "title": media.css("h3.p-tit a::text").extract_first(),
05             "publish-data": media.css("p.p-meta
span::text").extract_first(),
06         }
```

parse(response)是 Scrapy 的默认回调方法。也就是说，在生成的 Request 没有指定回调方法时，默认调用 parse()方法。

parse()方法主要负责处理 Response，返回抓取的数据，或者需要跟进的 URL。不只是 parse()方法，其他所有 Request 的回调方法都必须实现这些功能。

此方法以及任何其他 Request 回调方法必须返回可迭代的 Request、dicts 或 Item 对象。

4.1.4 Selector 选择器

在爬取网页时，最常做的任务就是从网页内容中提取数据。Scrapy 通过实现一套构建于 lxml 库上名为选择器(Selector)的机制来提取数据，主要通过特定的 Xpath 或者 CSS 表达式来选择 HTML 文件中的某个指定部分。

Selector 常用的方法主要有 4 个，分别说明如下。

- xpath()：传入 Xpath 表达式，返回表达式对应节点的选择器列表。
- css()：传入 CSS 表达式，返回表达式对应节点的选择器列表。
- extract()：以列表形式返回被选择元素的 Unicode 字符串。
- re()：返回通过正则表达式提取的 Unicode 字符串列表。

下面通过示例进行讲解。

测试页面结构如下：

```
01 <html>
02 <head>
03     <title>Test web site</title>
04  </head>
05  <body>
06     <p>line1</p>
07     <div id='images'>
08          <a href='/img/1' class='image'>name:image 1 <br />
                                         <img src='image1.jpg' /></a>
```

```
09          <a href='/img/2' class='image'>name:image 2 <br />
                                              <img src='image2.jpg' /></a>
10          <a href='/img/3' class='image'>name:image 3 <br />
                                              <img src='image3.jpg' /></a>
11          <a href='/img/4' class='image'>name:image 4 <br />
                                              <img src='image4.jpg' /></a>
12          <p>line2</p>
13      </div>
14    </body>
15 </html>
```

将其保存为example.html，在 Shell 终端中打开此文件：

```
scrapy shell D:\Scrapy\example.html
```

打开之后，就获得了一个 response 变量，并且在 response.selector 属性上附加了一个选择器。下面对选择器的使用进行介绍。

使用 xpath 方法获取 title 节点数据：

```
01 >>> response.xpath('//title/text()')
02 [<Selector (text) xpath=//title/text()>]
```

3 个>号表示在 Shell 模式中执行此命令，紧接着第 02 行为输出结果。

使用 css 方法获取 title 节点数据：

```
01 >>> response.css('title')
02 [<Selector (text) xpath=//title/text()>]
```

注意，通过.xpath()与.css()方法提取出来的都是类 SelectorList 实例的列表，我们可以通过extract()方法提取数据，获取的结果是一个列表：

```
01 >>> response.css('title::text').extract()
02 ['Test web site']
```

使用 xpath 方法获取 4 个 image 节点数据：

```
01 >>> response.xpath('//a[@class="image"]').extract()
02 ['<a href='/img/1' class='image'> name:image 1 <br />
      <img src='image1.jpg' /></a>', '<a href='/img/2' class='image'>
      name:image 2 <br /><img src='image2.jpg' /></a>', '<a href='/img/3'
       class='image'> name:image 3 <br /><img src='image3.jpg' /></a>', '<a
       href='/img/4' class='image'> name:image 4 <br /><img src='image4.jpg'
       /></a>']
```

获取第一条数据可以使用 extract_first()方法：

```
01 >>> response.xpath('//a[@class="image"]').extract_first()
02 '<a href='/img/1' class='image'> image 1 <br />
                                        <img src='image1.jpg' /></a>'
```

在使用xpath方法时，要特别注意相对路径的问题。在HTML示例文件中，div标签外有一个p标签，div内部也有一个p标签，如果以'/'开始的XPath语法定位标签，就说明与当前的Sclector无关，而是从整个文档开始定位，正确的使用方法是以'. '开始：

```
>>>div_tag = response.css('#images')
>>>div_tag.xpath('//p').extract_first()
'<p>line1</p>'
>>>div_tag.xpath('.//p').extract_first()
'<p>line2</p>'
```

在使用选择器的时候，有许多方法可以提取属性和文本信息。下面是一些应用举例：

```
>>> response.css('a::attr(href)').extract()
['/img/1', '/img/2', '/img/3', '/img/4']

>>> response.xpath('//a[contains(@href, "img")]/@href').extract()
['/img/1', '/img/2', '/img/3', '/img/4']

>>> response.css('a[href*=img]::attr(href)').extract()
['/img/1', '/img/2', '/img/3', '/img/4']

>>> response.xpath('//a[contains(@href, "img")]/img/@src').extract()
['image1.jpg', 'image2.jpg', 'image3.jpg', 'image4.jpg']

>>> response.css('a[href*=img] img::attr(src)').extract()
['image1.jpg', 'image2.jpg', 'image3.jpg', 'image4.jpg']
```

选择器可以嵌套使用：

```
>>>images = response.css('#images')
>>>images
[<Selector xpath="descendant-or-self::*[@id = 'images']" data='<div 
id="images">\r\n        <a href="/img'>]

>>> image_1 = images.css('a[href*="1"]').extract_first()
>>>image_1
'<a href="/img/1" class="image">name:image 1 <br><img src="image1.jpg"></a>'
```

提取的数据可以通过.re()正则表达式方法进一步提取筛选：

```
01 >>>response.xpath('//a[@class="image"]/text()').re(r'name:(.*)')
02 ['image 1','image 1','image 1','image 1']
```

相应的，re_first()提取第一条数据：

```
01 >>>response.xpath('//a[@class="image"]/text()').re_first(r'name:(.*)')
02 'image 1'
```

以上就是 Scrapy 选择器的一些基本用法，读者要熟练地使用不同的方法定位到需求的元素，在进行爬虫工作时会有很大的帮助。

4.2 通用爬虫

为了更加方便地爬取数据，针对常用情形，Scrapy 提供了几种通用的爬虫。比如，基于制定的规则抓取一个网页所有的链接，从站点地图抓取或解析 XML/CSV 源。本节要介绍的通用爬虫包括 CrawlSpider、XMLFeedSpider、CSVFeedSpider、SitemapSpider。

4.2.1 CrawlSpider

CrawlSpider 是抓取网站常用的 Spider，它提供了一个通过制定一些规则来达到跟进链接的方便机制。对一般网站的爬取来说，可以通过修改 CrawlSpider 来完成任务。

CrawlSpider 中最常用也是最重要的就是 rules 属性。rules 是一个或一组 Rule 对象，必须写成 tuple 形式。

每一个 Rule 对象定义了对目标网站的爬取行为，如果有多个 Rule 对象匹配了同一个链接，就说明第一个 Rule 会生效。下面通过示例详细介绍 Rule 对象。

Rule 定义如下：

```
class scrapy.spiders.Rule(
    link_extractor,callback=None,cb_kwargs=None,
    follow=None,process_links=None,process_request=None)
```

（1）link_extrator 是一个 Link Extrator 对象，定义了需要从已爬取的页面中提取哪些需要继续跟进的链接，通过匹配正则表达式来达到这一目的。

（2）callback 是一个回调函数或者字符串。传递字符串时，通过字符串查找本类中对应的函数名调用该函数。回调函数接收 Response 作为第一参数，同时返回一个包含 Item 或者 Request 对象的列表。

Rule 对象中不能使用 parse 作为回调函数，原因是 CrawlSpider 默认使用 parse 来实现爬虫逻辑，如果复写了 parse，那么 CrawlSpider 将不能继续正确执行。

（3）cb_kwargs 是一个字典对象，包含传递给回调函数（callback 指定的函数）的参数。

（4）follow 参数是一个布尔值，为 true 或者 false。若为 True，则需要跟进依据此条规则从 Response 中提取的链接；若为 False，则不跟进。若 callback 指定为 None，则 follow 默认为 True，否则为 False。

（5）process_links 的主要功能为过滤。它是一个回调函数或者字符串，传递字符串时，通过字符串查找本类中对应的函数名调用该函数。当使用本条 Rule 中制定的 link_extrator 从 Reponse 中获取链接列表时会调用此方法。

（6）process_request 是一个回调函数或者字符串，传递字符串时，通过字符串查找本类中对应的函数名调用该函数。通过本条 Rule 提取 Request 时调用此回调函数，该函数必须返回一个 Request 或者 None。

4.2.2 XMLFeedSpider

XMLFeedSpider 主要用于 RSS 源订阅内容的抓取。RSS 源是基于 XML 的一种信息聚合技术。该爬虫通过指定节点遍历达到抓取数据的目的。

以伯乐在线资讯类栏目的 RSS 订阅为例简单介绍 RSS 结构。单击订阅源按钮，如图 4.1 所示。

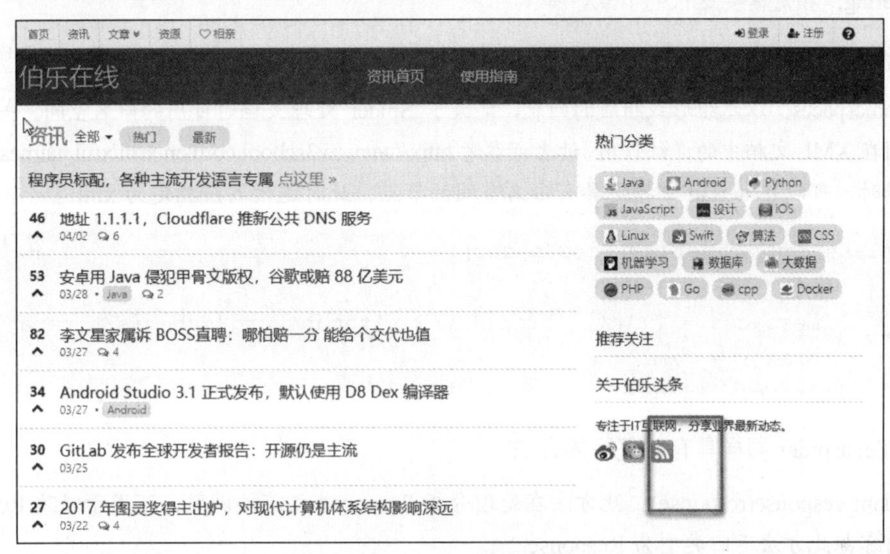

图 4.1 伯乐在线订阅

单击之后可以查看订阅内容源文件，部分如图 4.2 所示。

```
<item>
<title>C++委员会决定在C++20中弃用原始指针</title>
<link>http://top.jobbole.com/38659/</link>
<comments>http://top.jobbole.com/38659/#respond</comments>
<pubDate>Mon, 02 Apr 2018 10:48:28 +0000</pubDate>
<dc:creator><![CDATA[恒心]]></dc:creator>
        <category><![CDATA[业界]]></category>
<category><![CDATA[指针]]></category>

<guid isPermaLink="false">http://top.jobbole.com/?p=38659</guid>
<description><![CDATA[<p>会议上，除了讨论模块，概念，范围等最大事情之外，C ++ 委员会还讨论出了一个令人瞩目的特性：弃用原始指针！</p><p>The post <a href="http://top.jobbole.com/38659/">C++委员会决定在C++20中弃用原始指针</a> appeared first on <a href="http://top.jobbole.com">资讯 - 伯乐在线</a>.</p>]]>
</description>
<wfw:commentRss>http://top.jobbole.com/38659/feed/</wfw:commentRss>
<slash:comments>0</slash:comments>
</item>
<item>
<title>地址 1.1.1.1，Cloudflare 推新公共 DNS 服务</title>
<link>http://top.jobbole.com/38654/</link>
<comments>http://top.jobbole.com/38654/#comments</comments>
<pubDate>Mon, 02 Apr 2018 08:11:46 +0000</pubDate>
<dc:creator><![CDATA[伯小乐]]></dc:creator>
        <category><![CDATA[业界]]></category>
<category><![CDATA[CloudFlare]]></category>
<category><![CDATA[dns]]></category>

<guid isPermaLink="false">http://top.jobbole.com/?p=38654</guid>
<description><![CDATA[<p>4月1日，Cloudflare 宣布正式推出 1.1.1.1 公共 DNS 服务，号称任何人都可以使用它可以加快互联网访问速度并并保持连接私密性。Cloudflare 声称它将是“互联网上速度最快，隐私优先的消费者 DNS 服务”，此前类似的免费公共服务 OpenDNS 与 Google DNS 都已经服役了很长时间。</p><p>The post <a href="http://top.jobbole.com/38654/">地址 1.1.1.1，Cloudflare 推新公共 DNS 服务</a> appeared first on <a href="http://top.jobbole.com">资讯 - 伯乐在线</a>.</p>]]>
</description>
<wfw:commentRss>http://top.jobbole.com/38654/feed/</wfw:commentRss>
<slash:comments>6</slash:comments>
</item>
```

图 4.2 RSS 结构

可以看到，<item>...</item>标签之间是文章的一些相关信息，也就是需要抓取的内容，item称为一个 node（节点）。可以通过不同的节点获取我们需要抓取的内容。

XMLFeedSpider 中的一些属性如下：

- iterator：指定迭代器，迭代器主要用于分析数据 RSS 订阅源。可用的迭代器如下。
  - iternodes：性能高，基于正则表达式，是 XMLFEEDSpider 默认的迭代器。
  - html：使用 Selector 加载所有 DOM 结构进行分析，当数据量很大时会产生性能问题，优点是处理不合理标签时比较有用。
- xml：同 html 一样使用 Selector 进行分析，同样需要加载所有 DOM 结构，会产生性能问题。
- itertag：指定需要迭代的节点，如：

```
itertag = 'item'
```

- namespaces：以元组形式组成的列表，定义了 Spider 处理文档时可用的命名空间。关于命名空间在 XML 文档中的详细作用，读者可参考 http://www.w3school.com.cn/xml/xml_namespaces.asp。同时，可以在 itertag 中指定具有命名空间的节点，此时迭代器应指定为 xml：

```
01 class YourSpider(XMLFeedSpider):
02
03     namespaces = [('n', 'http://www.sitemaps.org/schemas/sitemap/0.9')]
04     iterator = 'xml'
05     itertag = 'n:url'
```

XMLFeedspider 同样具有可复写的方法：

- adapt_response(response)：此方法在处理分析 Response 之前被调用，可用于修改 Response 的内容。此方法返回类型为 Response。
- parse_node(response,selector)：当匹配到节点的时候，调用此方法进行数据处理。很重要的一点是，此方法必须复写，否则爬虫不会正常工作。该方法必须返回一个 Item、Request，或者一个包含 Item 或 Request 的迭代器。
- process_result(response,result)：当爬虫返回抓取结果时调用此方法。多用于在抓取结果传递给框架核心处理前做最后的修改。该方法必须接收一个结果列表和产生这些结果的 Response，返回一个包含 Item 或 Request 的结果列表。

4.2.3 CSVFeedSpider

CSVFeedSpider 与 XMLFeedspider 非常相似，区别是 XMLFeedspider 是根据节点来迭代数据的，而 CSVFeedSpider 是每行迭代。类似的，每行迭代调用的是 parse_row()方法。常用的属性方法如下：

- delimiter：字段分隔符，默认是英文逗号','。
- quotechar：CSV 字段中如果包含回车、引号、逗号，那么此字段必须用双引号引起来。此属性默认值为半角双引号。
- headers：CSV 文件的标题头，该属性值是一个列表。

- parse_row(response,row)：对每一行数据进行处理，接收由一个 Response、一个文件标题头组成的字典。

同 XMLFeedSpider 一样，在 CSVFeedSpider 中也可以复写 adapt_response 与 process_result 方法。

4.2.4 SitemapSpider

SitemapSpider 允许通过 Sitemap 发现 URL 链接来爬取一个网站。简单来讲，Sitemap 包含网站所有网址以及每个网址的其他元数据，包括上次更新的时间、更改的频率以及相对于网站上其他网址的重要程度为何等。Sitemap 有 TXT、XML、HTML 格式，一般以 XML 形式展现，代码如下：

```
<urlset xmlns="http://www.sitemaps.org/schemas/sitemap/0.9">
    <url>
        <loc>http://example.com/</loc>
        <lastmod>2016-09-06T00:00:16+08:00</lastmod>
        <changefreq>daily</changefreq>
        <priority>1.0</priority>
    </url>
    <url>
        <loc>http:// example.com/link.html</loc>
        <lastmod>2016-09-06T00:00:16+08:00</lastmod>
        <changefreq>daily</changefreq>
        <priority>0.8</priority>
    </url>
</urlset>
```

- loc 表示完整网址。
- lastmod 表示本网页最后修改时间。
- changefreq 表示更新频率。
- priority 用来指定此链接相对于其他链接的优先权比值。

SitemapSpider 常用属性如下。

- sitemap_urls：一个包含待爬取 url 的 sitemap 列表，也可以指定为 rebots.txt，表示从 rebots.txt 中提取 url。
- sitemap_rules：一个元祖列表，形如(regex,callback)，其中：
  - ♦ regex 表示需要从 sitmap 中提取的 url 的正则表达式，可以是一个字符串或者正则表达式对象。
  - ♦ callback 是处理对应的 url 的回调方法，如 sitemap_url = [('/price/','parse_price')]，提取到类似****/price/**链接时调用 parse_price 方法处理。需要注意的是，相同的链接只会调用第一个方法处理，并且如果此属性没有指定，那所有的链接默认使用 parse 方法处理。

- sitemap_follow：一个指定需要跟进的 sitemap 的正则表达式列表。当使用 Sitemap index files 来指向其他 sitemap 文件的站点时此属性有效。默认情况下，所有的 sitemap 都会被跟进。
- sitemap_alternate_links：指定当一个 url 有可选的链接时是否跟进。有些网站 url 块内会提供备用网址，如：

```
return item
<url>
<loc>http://example.com/</loc>
<xhtml:link rel="alternate" hreflang="de" href="http://example.com/de"/>
</url>
```

当启用 sitemap_alternate_links 属性时，两个网址都会被跟进；当不启用 sitemap_alternate_links 时，只会跟进 http://example.com/，此属性默认不启用。

4.3 爬虫实战

4.3.1 实战 1：CrawlSpider 爬取名人名言

【示例 4-1】使用 CrawlSpider 爬虫抓取数据

爬取名人名言（http://quotes.toscrape.com/）数据的流程为：首先，进入主页面（见图 4.3），提取名言内容、作者姓名、标签，然后通过作者链接跟进作者介绍页面（见图 4.4），提取作者相关信息。

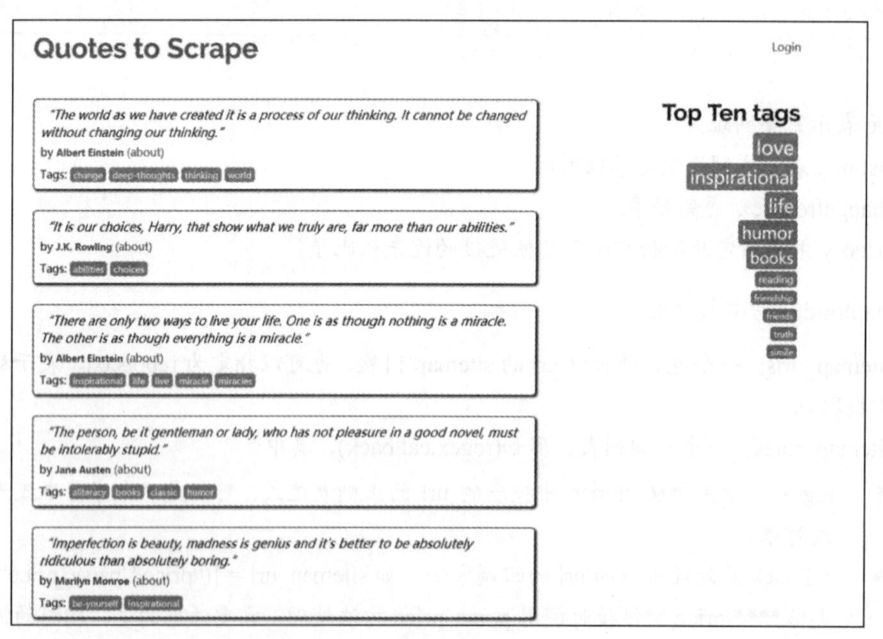

图 4.3　目标网站主页

Quotes to Scrape　　Login

Albert Einstein

Born: March 14, 1879 in Ulm, Germany

Description:

In 1879, Albert Einstein was born in Ulm, Germany. He completed his Ph.D. at the University of Zurich by 1909. His 1905 paper explaining the photoelectric effect, the basis of electronics, earned him the Nobel Prize in 1921. His first paper on Special Relativity Theory, also published in 1905, changed the world. After the rise of the Nazi party, Einstein made Princeton his permanent home, becoming a U.S. citizen in 1940. Einstein, a pacifist during World War I, stayed a firm proponent of social justice and responsibility. He chaired the Emergency Committee of Atomic Scientists, which organized to alert the public to the dangers of atomic warfare.At a symposium, he advised: "In their struggle for the ethical good, teachers of religion must have the stature to give up the doctrine of a personal God, that is, give up that source of fear and hope which in the past placed such vast power in the hands of priests. In their labors they will have to avail themselves of those forces which are capable of cultivating the Good, the True, and the Beautiful in humanity itself. This is, to be sure a more difficult but an incomparably more worthy task . . . " ("Science, Philosophy and Religion, A Symposium," published by the Conference on Science, Philosophy and Religion in their Relation to the Democratic Way of Life, Inc., New York, 1941). In a letter to philosopher Eric Gutkind, dated Jan. 3, 1954, Einstein stated: "The word god is for me nothing more than the expression and product of human weaknesses, the Bible a collection of honorable, but still primitive legends which are nevertheless pretty childish. No interpretation no matter how subtle can (for me) change this," (The Guardian, "Childish superstition: Einstein's letter makes view of religion relatively clear," by James Randerson, May 13, 2008). D. 1955.While best known for his mass-energy equivalence formula E = mc2 (which has been dubbed "the world's most famous equation"), he received the 1921 Nobel Prize in Physics "for his services to theoretical physics, and especially for his discovery of the law of the photoelectric effect". The latter was pivotal in establishing quantum theory.Einstein thought that Newtonion mechanics was no longer enough to reconcile the laws of classical mechanics with the laws of the electromagnetic field. This led to the development of his special theory of relativity. He realized, however, that the principle of relativity could also be extended to gravitational fields, and with his subsequent theory of gravitation in 1916, he published a paper on the general theory of relativity. He continued to deal with problems of statistical mechanics and quantum theory, which led to his explanations of particle theory and the motion of molecules. He also investigated the thermal properties of light which laid the foundation of the photon theory of light.He was visiting the United States when Adolf Hitler came to power in 1933 and did not go back to Germany. On the eve of World War II, he endorsed a letter to President Franklin D. Roosevelt alerting him to the potential development of "extremely powerful bombs of a new type" and recommending that the U.S. begin similar research. This eventually led to what would become the Manhattan Project. Einstein supported defending the Allied forces, but largely denounced the idea of using the newly discovered nuclear fission as a weapon. Later, with Bertrand Russell, Einstein signed the Russell-Einstein Manifesto, which highlighted the danger of nuclear weapons. Einstein was affiliated with the Institute for Advanced Study in Princeton, New Jersey, until his death in 1955.His great intellectual achievements and originality have made the word "Einstein" synonymous with genius.More:

图4.4　作者详情页

步骤如下：

步骤01 创建爬虫项目，如图4.5所示。

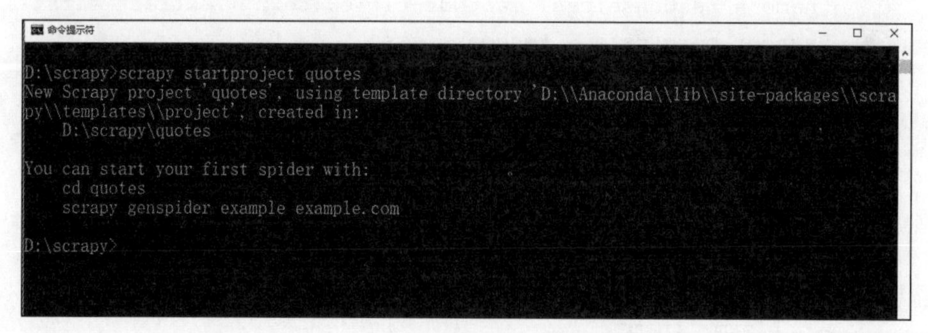

图4.5　创建Scrapy项目

步骤02 创建之后，在spiders文件夹下创建爬虫文件quotes.py，内容如下：

```
01 import scrapy
02 from scrapy.spider import CrawlSpider,Rule
03 from scrapy.linkextractors import LinkExtractor
04
05
06 class Quotes(CrawlSpider):
07     # 爬虫名称
08     name = "quotes"
```

```
09      allow_domain = ['quotes.toscrape.com']
10      start_urls = ['http://quotes.toscrape.com/']
11
12      # 设定规则
13      rules = (
14          # 对于 quotes 内容页 URL，调用 parse_quotes 处理
15          # 并以此规则跟进获取的链接
16          Rule(LinkExtractor(allow='/page/\d+'),callback='parse_quotes',
                                           follow=True),
17          # 对于 author 内容页 URL，调用 parse_author 处理，提取数据
18          Rule(LinkExtractor(allow='/author/\w+'),
                                           callback='parse_author')
19      )
20
21      # 提取内容页数据方法
22      def parse_quotes(self,response):
23          for quote in response.css(".quote"):
24              yield {
25                  'content': quote.css('.text::text').extract_first(),
26                  'author': quote.css('.author::text').extract_first(),
27                  'tags': quote.css('.tag::text').extract()
28              }
29      # 获取作者数据方法
30      def parse_author(self,response):
31          name = response.css('.author-title::text').extract_first()
32          author_born_date = response.css
                        ('.author-born-date::text').extract_first()
33          author_bron_location = response.css
                        ('.author-born-location::text').extract_first()
34          author_description = response.css
                        ('.author-description::text').extract_first()
35
36          return ({
37              'name':name,
38              'author_bron_date':author_born_date,
39              'author_bron_location':author_bron_location,
40              'author_description': author_description
41          })
```

步骤03 运行爬虫，执行 scrapy crawl quotes，如图 4.6 所示。

内容提取结果如图 4.7 所示。

图4.6　运行爬虫

图4.7　内容提取结果

作者信息提取结果如图4.8所示。

图4.8　作者信息提取结果

这里主要对 rules 中制定的规则进行讲解。

（1）Rule(LinkExtractor(allow='/page/\d+'), callback='parse_quotes', follow=True)

查看页面元素，查看下一页按钮，链接地址为/page/+页面数字，如图 4.9 所示。

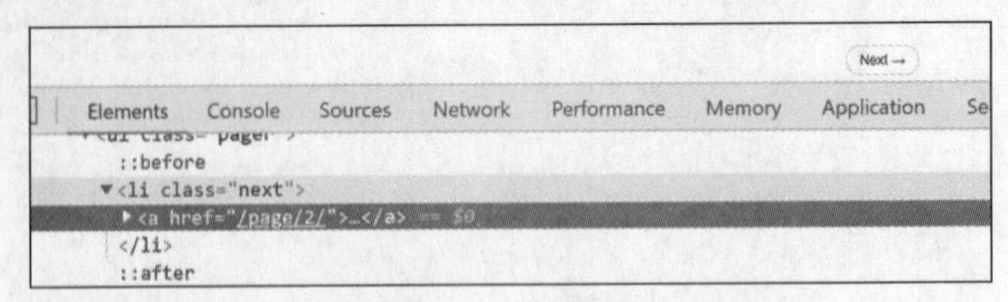

图 4.9　内容页 URL

所以第 1 条 Rule 中，爬取符合'/page/\d+'正则表达式的所有链接即为所有内容页。然后根据 callback 调用 parse_quotes 处理，提取相关数据，由于 follow=True，因此跟进 Response 返回的所有符合规则的链接，也就是内容页的链接。

（2）Rule(LinkExtractor(allow='/author/\w+'), callback='parse_author')

如图 4.10 所示，作者介绍页的链接地址为/author/+作者姓名。

```
▸<span class="text" itemprop="text">…</span>
▾<span>
   "by "
   <small class="author" itemprop="author">Albert Einstein</small>
   <a href="/author/Albert-Einstein">(about)</a> == $0
 </span>
▸<div class="tags">…</div>
</div>
```

图 4.10　作者介绍页 URL

因此，在第 2 条 Rule 中，对所有满足正则表达式'/author/\w+'的链接进行抓取，即可获取作者介绍页的内容，调用 parse_author 进行数据提取。

本例中使用.css()方法进行数据提取，读者可以尝试使用.xpath()进行数据提取练习。

4.3.2　实战 2：XMLFeedSpider 爬取伯乐在线的 RSS

【示例 4.2】一个使用 XMLFeedspider 抓取伯乐在线 RSS 订阅信息的爬虫实例

爬取伯乐在线资讯文章，内容和结构分别如图 4.1 和图 4.2 所示。

（1）创建爬虫项目，如图 4.11 所示。

```
命令提示符
D:\Scrapy>scrapy startproject xmlfeedspider
New Scrapy project 'xmlfeedspider', using template directory 'd:\\software\\
anaconda\\lib\\site-packages\\scrapy\\templates\\project', created in:
    D:\Scrapy\xmlfeedspider

You can start your first spider with:
    cd xmlfeedspider
    scrapy genspider example example.com

D:\Scrapy>
```

图 4.11　创建项目

（2）使用XMLFeedSpider模板创建爬虫，如图4.12所示。

```
D:\Scrapy\xmlfeedspider>scrapy genspider -t xmlfeed jobbole jobbole.com
Created spider 'jobbole' using template 'xmlfeed' in module:
  xmlfeedspider.spiders.jobbole

D:\Scrapy\xmlfeedspider>
```

图4.12 使用XMLFeedspider模板创建爬虫

（3）这次我们使用Item收集数据，在xmlfeedspider文件夹中，修改items.py文件，代码如下：

```
01 import scrapy
02
03
04 class JobboleItem(scrapy.Item):
05     # define the fields for your item here like:
06     # name = scrapy.Field()
07     # 文章标题
08     title = scrapy.Field()
09     # 发表日期
10     public_date = scrapy.Field()
11     # 文章链接
12     link = scrapy.Field()
```

（4）在spiders文件中，修改jobble.py文件，代码如下：

```
01 from scrapy.spiders import XMLFeedSpider
02 # 导入item
03 from xmlfeedspider.items import XmlfeedspiderItem
04
05 class JobboleSpider(XMLFeedSpider):
06     name = 'jobbole'
07     allowed_domains = ['jobbole.com']
08     start_urls = ['http://top.jobbole.com/feed/']
09     iterator = 'iternodes'          # 迭代器，不指定的话，默认是iternodes
10     itertag = 'item'                # 抓取item节点
11
12     def parse_node(self, response, selector):
13         item = XmlfeedspiderItem()
14         item['title'] = selector.css('title::text').extract_first()
15         item['public_date'] = selector.css
                                  ('pubDate::text').extract_first()
16         item['link'] = selector.css('link::text').extract_first()
17         return item
```

（5）在 setting 中，需修改的配置如下：

```
ROBOTSTXT_OBEY = False            # 不依据 robots.txt 规则
```

（6）运行爬虫，查看结果，如图 4.13 所示。

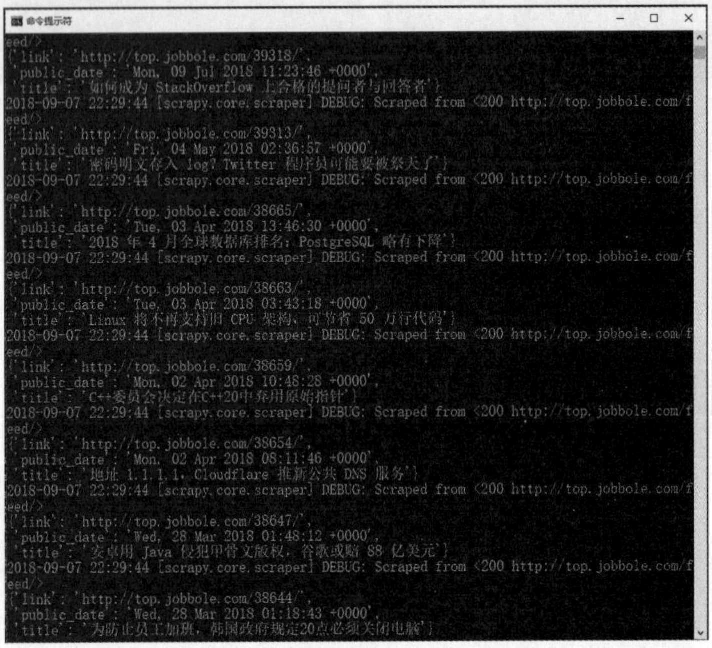

图 4.13　XMLFeedspider 运行结果

可以看到，根据指定的节点抓取了所有文章的信息。

4.3.3　实战 3：CSVFeedSpider 提取 csv 文件数据

【示例 4-3】使用 CSVFeedSPider 提取 csv 文件数据

从贵州省数据开放平台下载科技特派员 csv 文件，文件地址为：

http://gzopen.oss-cn-guizhou-a.aliyuncs.com/科技特派员.csv

（1）使用命令创建项目：

```
>>>scrapy startproject csvfeedspider
```

（2）进入项目目录：

```
>>>cd csvfeedspider
>>> scrapy genspider -t csvfeed csvdata gzdata.gov.cn
```

（3）编写 items.py 文件，代码如下：

```
01  import scrapy
02
```

```
03
04 class CsvspiderItem(scrapy.Item):
05     # define the fields for your item here like:
06     # 姓名
07     name = scrapy.Field()
08     # 研究领域
09     SearchField = scrapy.Field()
10     # 服务分类
11     Service = scrapy.Field()
12     # 专业特长
13     Specialty = scrapy.Field()
```

（4）编写爬虫文件 csvdata.py，内容如下：

```
01 # -*- coding: utf-8 -*-
02 from scrapy.spiders import CSVFeedSpider
03 from csvfeedspider.items import CsvspiderItem
04
05
06 class CsvparseSpider(CSVFeedSpider):
07     name = 'csvdata'
08     allowed_domains = ['gzdata.gov.cn']
09     start_urls = ['http://gzopen.oss-cn-guizhou-a.aliyuncs.com
                                                        /科技特派员.csv']
10     headers = ['name', 'SearchField', 'Service', 'Specialty']
11     delimiter = ','
12     quotechar = "\n"
13
14     # Do any adaptations you need here
15     def adapt_response(self, response):
16         return response.body.decode('gb18030')
17
18     def parse_row(self, response, row):
19         i = CsvspiderItem()
20         i['name'] = row['name']
21         i['SearchField'] = row['SearchField']
22         i['Service'] = row['Service']
23         i['Specialty'] = row['Specialty']
24         return i
```

在 adapt_response()方法中，我们对 response 做了编码处理，使之能正常地提取中文数据，如果不做处理，提取出来的将是乱码数据。

（5）运行爬虫

```
>>>scrapy crawl csvdata
```

查看运行结果，如图 4.14 所示。

```
2019-03-06 21:44:43 [scrapy.core.scraper] DEBUG: Scraped from <200 http://gzopen.oss-cn-guizhou-a.aliyuncs.com/%E7%A7%91%E6%8A%80%E7%89%B9%E6%B4%BE%E5%91
%98.csv>
{'SearchField': '生态学', 'Service': '技术咨询', 'Specialty': '科长', 'name': '汪建文'}
2019-03-06 21:44:43 [scrapy.core.scraper] DEBUG: Scraped from <200 http://gzopen.oss-cn-guizhou-a.aliyuncs.com/%E7%A7%91%E6%8A%80%E7%89%B9%E6%B4%BE%E5%91
%98.csv>
{'SearchField': '动物医学',
 'Service': '实用技术推广及培训',
 'Specialty': '高级兽医师',
 'name': '刘万勇'}
2019-03-06 21:44:43 [scrapy.core.scraper] DEBUG: Scraped from <200 http://gzopen.oss-cn-guizhou-a.aliyuncs.com/%E7%A7%91%E6%8A%80%E7%89%B9%E6%B4%BE%E5%91
%98.csv>
{'SearchField': '生物产业、观光农业、中药材生产……',
 'Service': '生物产业、观光农业……',
 'Specialty': '副院长',
 'name': '张明生'}
```

图 4.14 csvdata 运行结果

4.3.4 实战 4：SitemapSpider 爬取博客园文章

【示例 4-4】使用 SitemapSpider 爬取博客园 cnblogs.com 文章

博客园的 Sitemap 地址为 http://www.cnblogs.com/sitemap.xml，查看其中内容，如图 4.15 所示。我们需要抓取 python 类别下的所有文章的标题、作者、文章网址。

```
← → C ⌂ ⓘ 不安全 | www.cnblogs.com/sitemap.xml
    <lastmod>2019-03-02T16:46:23+00:00</lastmod>
    <changefreq>hourly</changefreq>
    <priority>0.8</priority>
  </url>
▼<url>
    <loc>http://www.cnblogs.com/cate/python/</loc>
    <lastmod>2019-03-02T16:46:23+00:00</lastmod>
    <changefreq>hourly</changefreq>
    <priority>0.8</priority>
  </url>
▼<url>
    <loc>http://www.cnblogs.com/cate/codelife/</loc>
    <lastmod>2019-03-02T16:46:23+00:00</lastmod>
    <changefreq>hourly</changefreq>
    <priority>0.8</priority>
  </url>
▼<url>
    <loc>http://www.cnblogs.com/cate/job/</loc>
    <lastmod>2019-03-02T16:46:23+00:00</lastmod>
    <changefreq>hourly</changefreq>
    <priority>0.8</priority>
  </url>
```

图 4.15 cnblogs 网站 Sitemap

python 类别下的文章如图 4.16 所示，查看文章标题、url、作者元素定位信息。

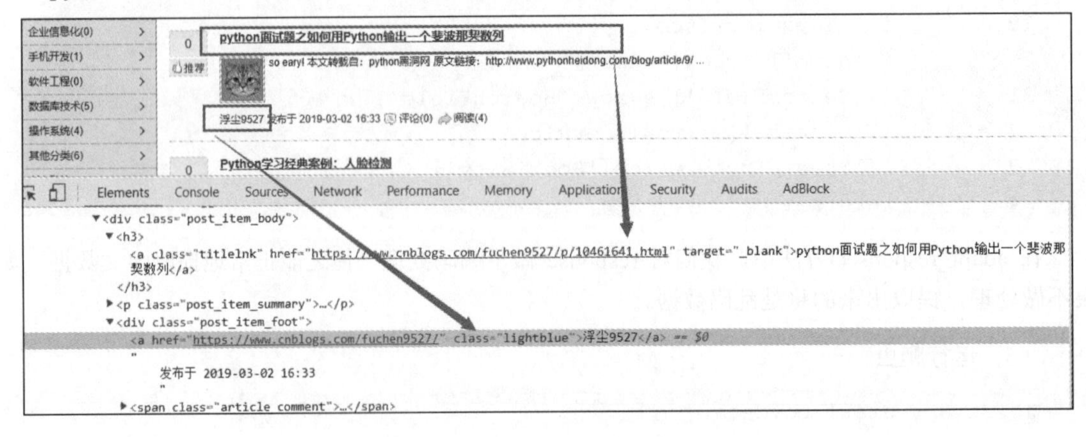

图 4.16 python 类别下的文章元素定位信息

分析完之后，开始创建项目，编写代码。

（1）创建项目：

```
>>>scrapy startproject cnblogs
```

（2）修改 items.py 文件，抓取文章标题、url、作者，内容如下：

```
01 import scrapy
02
03
04 class CnblogsItem(scrapy.Item):
05     # define the fields for your item here like:
06     # 文章标题
07     title = scrapy.Field()
08     # 文章 url
09     url = scrapy.Field()
10     # 文章作者
11     author = scrapy.Field()
12
```

（3）在 spiders 文件夹内创建 articles.py 爬虫文件，内容如下：

```
01 from scrapy.spiders import SitemapSpider
02 from cnblogs.items import CnblogsItem
03
04 class MySpider(SitemapSpider):
05     name = 'articles'
06     # Sitemap 地址
07     sitemap_urls = ['http://www.cnblogs.com/sitemap.xml']
08     # 从 Sitemap 中提取 url 的规则，并指定回调方法
09     sitemap_rules = [
10         # 抓取 ***/cate/python/**的 url，调用 parse_python 处理
11         ('/cate/python/','parse_python')
12     ]
13
14     # 回调方法
15     def parse_python(self,response):
16         articles = response.css('.post_item')
17
18         for article in articles:
19             item = CnblogsItem()
20             # 文章标题
21             item['title'] = article.css
                            ('.titlelnk::text').extract_first()
22             # 文章 url
23             item['url'] = article.css
                            ('.titlelnk::attr(href)').extract_first()
```

```
    24            # 文章作者
    25            item['author'] =
article.css('.lightblue::text').extract_first()
    26            yield item
```

使用 sitemap_rules 定义提取 url 的规则，并指定回调方法，保存到 CnblogsItem 中。

（4）运行爬虫：

```
scrapy crawl articles
```

查看输出结果，如图 4.17 所示。

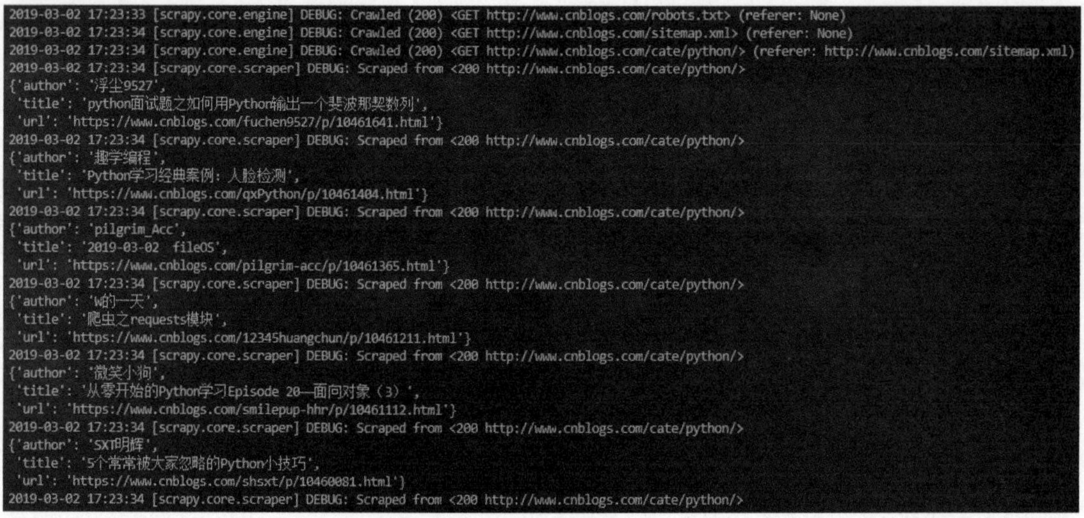

```
2019-03-02 17:23:33 [scrapy.core.engine] DEBUG: Crawled (200) <GET http://www.cnblogs.com/robots.txt> (referer: None)
2019-03-02 17:23:34 [scrapy.core.engine] DEBUG: Crawled (200) <GET http://www.cnblogs.com/sitemap.xml> (referer: None)
2019-03-02 17:23:34 [scrapy.core.engine] DEBUG: Crawled (200) <GET http://www.cnblogs.com/cate/python/> (referer: http://www.cnblogs.com/sitemap.xml)
2019-03-02 17:23:34 [scrapy.core.scraper] DEBUG: Scraped from <200 http://www.cnblogs.com/cate/python/>
{'author': '浮尘9527',
 'title': 'python面试题之如何用Python输出一个斐波那契数列',
 'url': 'https://www.cnblogs.com/fuchen9527/p/10461641.html'}
2019-03-02 17:23:34 [scrapy.core.scraper] DEBUG: Scraped from <200 http://www.cnblogs.com/cate/python/>
{'author': '趣学编程',
 'title': 'Python学习经典案例：人脸检测',
 'url': 'https://www.cnblogs.com/qxPython/p/10461404.html'}
2019-03-02 17:23:34 [scrapy.core.scraper] DEBUG: Scraped from <200 http://www.cnblogs.com/cate/python/>
{'author': 'pilgrim_Acc',
 'title': '2019-03-02  fileOS',
 'url': 'https://www.cnblogs.com/pilgrim-acc/p/10461365.html'}
2019-03-02 17:23:34 [scrapy.core.scraper] DEBUG: Scraped from <200 http://www.cnblogs.com/cate/python/>
{'author': 'W的一天',
 'title': '爬虫之requests模块',
 'url': 'https://www.cnblogs.com/12345huangchun/p/10461211.html'}
2019-03-02 17:23:34 [scrapy.core.scraper] DEBUG: Scraped from <200 http://www.cnblogs.com/cate/python/>
{'author': '微笑小狗',
 'title': '从零开始的Python学习Episode 20—面向对象（3）',
 'url': 'https://www.cnblogs.com/smilepup-hhr/p/10461112.html'}
2019-03-02 17:23:34 [scrapy.core.scraper] DEBUG: Scraped from <200 http://www.cnblogs.com/cate/python/>
{'author': 'SXT明辉',
 'title': '5个常常被大家忽略的Python小技巧',
 'url': 'https://www.cnblogs.com/shsxt/p/10460081.html'}
2019-03-02 17:23:34 [scrapy.core.scraper] DEBUG: Scraped from <200 http://www.cnblogs.com/cate/python/>
```

图 4.17　抓取 cnblogs 运行结果

第5章

Scrapy 管道

在 Scrapy 提供的各种组件中，管道是除了爬虫以外我们接触最多的组件。管道可以为我们提供各种处理数据的能力，比如清洗保存数据、处理图片与文件，还可以在管道中进行数据存储操作，存入数据库或存储到本地文件等。

本章主要的知识点有：

- 管道的编写方法
- 文件处理
- 图片处理
- 数据库存储

5.1 管道简介

管道（Item Pipeline）的主要作用是处理抓取的数据。当爬虫抓取到数据并转化为 Item 之后，会传递给 Item Pipeline 做进一步处理，包括：

- 清洗数据
- 检查抓取的数据是否有效
- 去重
- 保存数据

一个项目可以包含多个管道，通过爬虫收集到的 Item 会依次按指定顺序传递给管道进行处理。每一个管道都是实现一些指定方法的 Python 类。这些方法接收 Item 作为参数，判断此 Item 是否进行下一步处理，还是丢弃。

5.2 编写自定义管道

管道的编写很简单，每一个 Item Pipeline 都是独立的 Python 类，只要实现一些指定的方法即可实现管道的定制：

- process_item(self, item, spider)：process_item()是每个管道都必须实现的方法。数据处理工作都在此方法中进行，该方法返回一个有数据的 dict、Item（或者继承类），返回 Twisted Deferred 或抛出 DropItem 异常，丢弃的数据不会再传递到其他管道进行处理。其中，参数 item 对象是被爬取的 Item，参数 spider 代表爬取该 Item 的 Spider。
- open_spider(self, spider)：当爬虫开启的时候执行此方法，一般做一些初始化工作，比如连接数据库。
- close_spider(self, spider)：当爬虫关闭的时候执行此方法，可以做一些收尾工作、关闭数据库等。
- from_crawl(cls, crawler)：该方法是一个类方法。调用此方法会通过初始化 crawler 对象返回一个 Pipeline 实例。通过 crawler 对象可以返回 Scrapy 的所有核心组件，如全局配置信息。

【示例 5-1】以抓取并存储伯乐在线最新文章为例演示管道基本用法

（1）创建项目：

```
>>>scrapy startproject jobbole_article
```

（2）使用 genspider 命令创建爬虫文件：

```
>>>cd jobbole_article
>>>scrapy genspider article jobbole.com
```

（3）进入项目目录，修改 items.py 文件，内容如下：

```
01 import scrapy
02
03
04 class JobboleArticleItem(scrapy.Item):
05     # define the fields for your item here like:
06     # name = scrapy.Field()
07     # 文章标题
08     title = scrapy.Field()
09     # 内容摘要
10     summary = scrapy.Field()
11     # 发表日期
12     publish_date = scrapy.Field()
13     # 标签
14     tag = scrapy.Field()
```

（4）使用 JSON 格式化存储抓取数据，修改 pipelines.py，内容如下：

```
01 import json
02
03
04 class JobboleArticlePipeline(object):
05     # 当启动爬虫时，打开 items.json 文件，准备写入数据
06     def open_spider(self, spider):
07         self.file = open('items.json','w')
08
09     # 当爬虫执行结束时，关闭打开的文件
10     def close_spider(self, spider):
11         self.file.close()
12
13     # 将抓取到的数据做 json 序列化存储
14     def process_item(self, item, spider):
15         line = json.dumps(dict(item), ensure_ascii=False) + "\n"
16         self.file.write(line)
17         return item
```

在这里简单介绍一下 JSON（JavaScript Object Notation）。JSON 是一种轻量级的数据交换格式，表示出来就是一个字符串，可以被所有语言读取。在 Python 中，JSON 处理文件本质上就是一个编码、解码的过程。

JSON 库中的 dump 和 dumps 方法实现 JSON 编码功能。区别是 dump 方法将编码后的数据保存到文件中，而 dumps 方法产生一个 JSON 字符串，使用方法如下：

```
>>> student = {"name":"Ming", "age":14}
>>> import json
>>> student = {"name":"Ming","age":16}
>>> json.dumps(student)
'{"name": "Ming", "age": 16}'
>>> f = open('student.json','w')
>>> json.dump(student,f)
>>> f.close()
```

至于上面代码中使用的 ensure_ascii=False 参数，是因为 JSON 在处理中文时，默认使用的是 ASCII 编码，指定此参数为 False 才能正确地在文件中保存中文格式数据。

load 和 loads 方法实现 JSON 解码功能，load 需要从文件加载数据解码，而 loads 加载字符串进行解码，使用方法如下：

```
>>> books=[{'name':'Python','price':24},{'name':'Java','price':45},
{'name':'PHP','price':33}]
>>> json_str=json.dumps(books)
>>> json.loads(json_str)
[{'name': 'Python', 'price': 24}, {'name': 'Java', 'price': 45}, {'name':
'PHP', 'price': 33}]
```

```
>>> f=open('books.json','w')
>>> json.dump(books,f)
>>> f.close()
>>> json_file=open('books.json','r')
>>> json.load(json_file)
[{'name': 'Python', 'price': 24}, {'name': 'Java', 'price': 45}, {'name':
'PHP', 'price': 33}]
```

（5）编写完管道之后，需要启用管道，不然不会生效。只需要将管道添加到 settings.py 文件的 ITEM_PIPELINES 变量中：

```
# Configure item pipelines
# See https://doc.scrapy.org/en/latest/topics/item-pipeline.html

ITEM_PIPELINES = {
   'jobbole_article.pipelines.JobboleArticlePipeline': 300,
}
```

（6）最后进行爬虫文件的编写。在每一页中获取 Item 所需元素，然后查找是否有下一页，一直循环下去。spiders/article.py 中的代码如下：

```
01 # -*- coding: utf-8 -*-
02 import scrapy
03 from jobbole_article.items import JobboleArticleItem
04
05
06 class ArticleSpider(scrapy.Spider):
07     name = 'article'
08     allowed_domains = ['jobbole.com']
09     start_urls = ['http://blog.jobbole.com/all-posts/']
10
11     def parse(self, response):
12         all_post = response.css(".post")
13         for post in all_post:
14             item = JobboleArticleItem()
15             item['title'] = post.css('.archive-title::text').extract_first()
16             item['summary'] = post.css('.excerpt p::text').extract_first()
17             # 根据正则表达式提取发表日期
18             item['publish_date'] = post.css('.post-meta p::text').
                                          re_first(r'\d{4}/\d{2}/\d{2}')
19             # Tag 标签可能有多个，因此不需要获取第一个值，保存列表即可
20             item['tag'] = post.xpath(".//a[2]/text()").extract()
21             yield item
22
23         # 检查是否有下一页 url，如果有下一页，则调用 parse 进行处理
24         next_page = response.css('.next::attr(href)').extract_first()
25         if next_page:
```

```
26            yield scrapy.Request(next_page,callback=self.parse)
27
```

（7）运行爬虫：

```
scrapy crawl article
```

（8）查看生成的 item.json 文件，查看保存的数据，如图 5.1 所示。

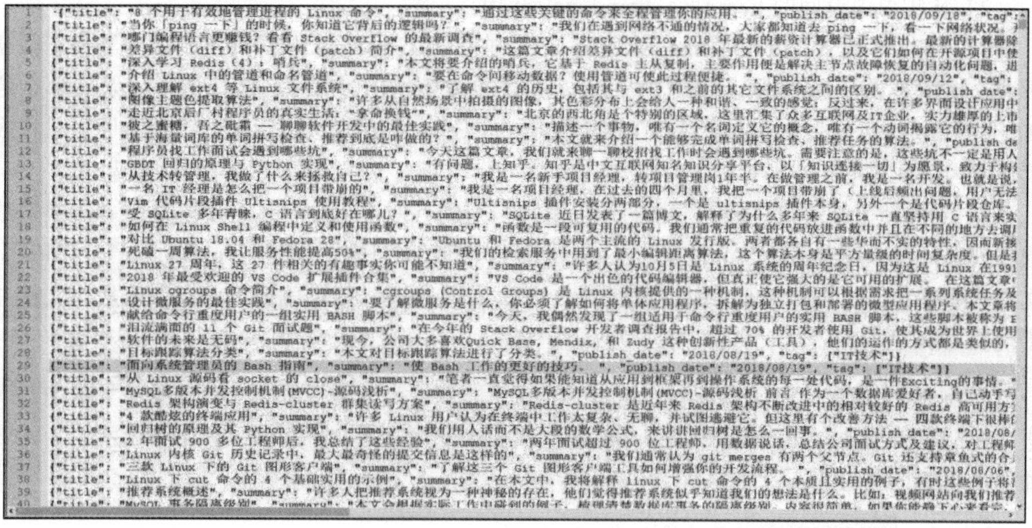

```
1  {"title": "8 个用于有效地管理进程的 Linux 命令", "summary": "通过这些关键的命令来全程管理你的应用。 ", "publish_date": "2018/09/18", "tag":
2  {"title": "当你「ping 一下」的时候，你知道它背后的逻辑吗？", "summary": "我们在遇到网络不通的情况，大家都知道去 ping 一下，看一下网络状况。
3  {"title": "哪门编程语言更赚钱？看看 Stack Overflow 的最新调查", "summary": "Stack Overflow 2018 年最新的薪资计算器已正式推出。最新的计算器
4  {"title": "差异文件（diff）和补丁文件（patch）简介", "summary": "这篇文章介绍差异文件（diff）和补丁文件（patch），以及它们如何在开源项目中使
5  {"title": "深入学习 Redis（4）：哨兵", "summary": "本文将要介绍的哨兵，它基于 Redis 主从复制，主要作用便是解决主节点故障恢复的自动化问题，进
6  {"title": "介绍 Linux 中的管道和命名管道", "summary": "要在命令间移动数据？使用管道可使此过程便捷。 ", "publish date": "2018/09/12", "tag":
7  {"title": "深入理解 ext4 等 Linux 文件系统", "summary": "了解 ext4 的历史，包括其与 ext3 和之前的其它文件系统之间的区别。 ", "publish date":
8  {"title": "图像主题色提取算法", "summary": "许多从自然场景中拍摄的图像，其色彩分布上会给人一种和谐、一致的感觉；反过来，在许多界面设计应用中
9  {"title": "走近北京后厂村程序员的真实生活："拿命换钱"", "summary": "北京的西北角是个特别的区域，这里汇集了众多互联网及IT企业，实力雄厚的上市
10 {"title": "彼之蜜糖，吾之砒霜 — 聊聊软件开发中的最佳实践", "summary": "描述一个事物，唯有一个名词定义它的概念，唯有一个动词揭露它的行为，唯
11 {"title": "基于海量词库的单词拼写检查、推荐到底是咋做的？", "summary": "本文就来介绍一个能够完成单词拼写检查，推荐任务的算法。", "publish da
12 {"title": "程序员找工作面试会遇到哪些坑", "summary": "今天这篇文章，我们就来聊一聊校招找工作时会遇到哪些坑。需要注意的是，这些坑不一定是用人
13 {"title": "GBDT 回归的原理与 Python 实现", "summary": "有问题，上知乎。知乎是中文互联网知名知识分享平台，以「知识连接一切」为愿景，致力于构
14 {"title": "从技术转管理，我做了什么来拯救自己？", "summary": "我是一名新手项目经理，转项目管理岗1年半。在做管理之前，我是一名开发。也就是说，
15 {"title": "一名 IT 经理是怎么把一个项目带崩的", "summary": "我是一名项目经理，在过去的四个月里，我把一个项目带崩了（上线后频出问题，用户无法
16 {"title": "Vim 代码片段插件 Ultisnips 使用教程", "summary": "Ultisnips 插件安装分两部分，一个是 ultisnips 插件本身，另外一个是代码片段仓库。
17 {"title": "受 SQLite 多年青睐，C 语言到底好在哪儿？", "summary": "SQLite 近日发表了一篇博文，解释了为什么多年来 SQLite 一直坚持用 C 语言来实
18 {"title": "如何在 Linux Shell 编程中定义和使用函数", "summary": "函数是一段可复用的代码。我们通常把重复的代码放进函数中并且在不同的地方去调
19 {"title": "对比 Ubuntu 18.04 和 Fedora 28", "summary": "Ubuntu 和 Fedora 是两个主流的 Linux 发行版。两者都各自有一些华而不实的特性，因而新接
20 {"title": "死磕一周算法，我让服务性能提高50%", "summary": "我们的检索服务中用到了最小编辑距离算法，这个算法本身是平方量级的时间复杂度。但是
21 {"title": "Linux 27 周年，这 27 件相关的有趣事实你可能不知道", "summary": "许多人认为10月5日是 Linux 系统的周年纪念日，因为这是 Linux 在1991
22 {"title": "2018 年最受欢迎的 VS Code 扩展插件合集", "summary": "VS Code 是一个出色的代码编辑器，但真正使它强大的是它可用的扩展。 在这篇文章
23 {"title": "Linux cgroups 命令简介", "summary": "cgroups (Control Groups) 是 Linux 内核提供的一种机制，这种机制可以根据需求把一系列系统任务及
24 {"title": "设计微服务的最佳实践", "summary": "要了解微服务是什么，你必须了解如何将单体应用程序，拆解为独立打包和部署的微型应用程序。本文章将
25 {"title": "献给命令行重度用户的一组实用 BASH 脚本", "summary": "今天，我偶然发现了一组适用于命令行重度用户的实用 BASH 脚本，这些脚本被称为 E
26 {"title": "泪流满面的 11 个 Git 面试题", "summary": "在今年的 Stack Overflow 开发者调查报告中，超过 70% 的开发者使用 Git，使其成为世界上使用
27 {"title": "软件的未来是无码", "summary": "现今，公司大多喜欢Quick Base, Mendix, 和 Zudy 这种创新性产品（工具），他们的运作的方式都是类似的，
28 {"title": "目标跟踪算法分类", "summary": "本文对目标跟踪算法进行了分类。", "publish_date": "2018/08/19", "tag": ["IT技术"]}
29 {"title": "面向系统管理员的 Bash 指南", "summary": "使 Bash 工作的更好的技巧。 ", "publish_date": "2018/08/19", "tag": ["IT技术"]}
30 {"title": "从 Linux 源码看 socket 的 close", "summary": "笔者一直觉得如果能知道从应用到框架再到操作系统的每一处代码，是一件Exciting的事情。"
31 {"title": "MySQL多版本并发控制机制(MVCC)-源码浅析", "summary": "MySQL多版本并发控制机制(MVCC)-源码浅析 前言 作为一个数据库爱好者，自己动手写
32 {"title": "Redis 架构演变与 Redis-cluster 群集读写方案", "summary": "Redis-cluster 是近年来 Redis 架构不断改进中的相对较好的 Redis 高可用方
33 {"title": "4 款酷炫的终端应用", "summary": "许多 Linux 用户认为在终端中工作太复杂、无聊，并试图逃避它。但这里有个改善方法 — 四款终端下很棒
34 {"title": "回归树的原理及其 Python 实现", "summary": "我们用人话而不是大段的数学公式，来讲讲回归树是怎么一回事。", "publish date": "2018/08/
35 {"title": "2 年面试 900 多位工程师后，我总结了这些经验", "summary": "两年面试超过 900 位工程师，用数据说话，总结公司面试方式及建议，对工程师
36 {"title": "Linux 内核 Git 历史记录中，最大最奇怪的提交信息是这样的", "summary": "我们通常认为 git merges 有两个父节点。Git 还支持章鱼式的合
37 {"title": "三款 Linux 下的 Git 图形客户端", "summary": "了解这三个 Git 图形客户端工具如何增强你的开发流程。 ", "publish date": "2018/08/06",
38 {"title": "Linux 下 cut 命令的 4 个基础实用的示例", "summary": "在本文中，我将解释 linux 下 cut 命令的 4 个本质且实用的例子，有时这些例子将
39 {"title": "推荐系统概述", "summary": "许多人把推荐系统视为一种神秘的存在，他们觉得推荐系统似乎知道我们的想法是什么。比如，视频网站向我们推荐
40 {"title": "MySQL 事务隔离级别", "summary": "本文会根据实际工作中碰到的例子，梳理清楚数据库事务的隔离级别，内容很简单，如果你能静下心来看完，
```


图 5.1　使用 JSON 保存数据

5.3　下载文件和图片

在使用爬虫获取数据时，有时不仅要抓取“字符串”，还要进行图片和文件的获取。比如我们抓取租房信息时，不仅要抓取地段、租金等信息，同时要把房屋图片一起抓下来。针对这些情况，Scrapy 提供了一些可重用的 Item Pipeline。这些管道有一些共同的方法和结构，通常叫作 Media Pipelines，而我们最常用的就是 FilesPipeline 和 ImagesPipeline，分别用于下载文件和图片。这两种管道都包含以下特性：

- 避免重复下载最近下载过的数据。
- 指定存储的位置，可使用本地文件系统或者云端存储。

同时，ImagesPipeline 还有一些额外特性：

- 将下载的图片转换成通用的 JPG 格式和 RGB 模式。
- 为下载的图片生成缩略图。
- 检查图片的宽/高，确保能够满足最小要求。

管道同时会为计划中下载的文件 URL 保存一个内部队列，与包含同样文件的 Response 相关联，从而避免重复下载几个 Item 共用的图片。

5.3.1 文件管道

在爬虫项目中，使用文件管道的工作流程如下：

（1）在爬虫中，抓取到 Item 并把期望的 URL 放入 file_urls 内。

（2）从爬虫内返回的 Item 进入 Item Pipeline 内。

（3）当 Item 进入 FilesPipeline 时，file_urls 内的 URL 将被 Scrapy 内置的调度器和下载器安排下载，这就是说，调度器和下载器中间件是可以重复使用的。如果是更高的优先级，那么这些 URL 会在抓取其他页面之前被处理。在这个管道内的 Item 会保持锁定状态，直到文件下载完成或因其他原因下载失败。

（4）当文件下载完成之后，下载的结果将会填充到 fiels 字段中。这个字段是由 dict 类型数据组成的列表，包含下载路径、源地址（从 file_urls 中获取）、图片校验码（checksum）。files 字段中文件的顺序与 file_urls 文件 URL 顺序保持一致。如果下载失败，就会记录错误信息，文件并不会出现在 files 字段中。

【示例 5-2】清楚文件下载流程之后，下面以抓取深圳市人力资源和社会保障局下载中心人才引进（http://hrss.sz.gov.cn/wsbs/xzzx/rcyj/）专栏页面中的第一页文件为例（如图 5.2 所示），看一下如何使用 FilesPipeline。

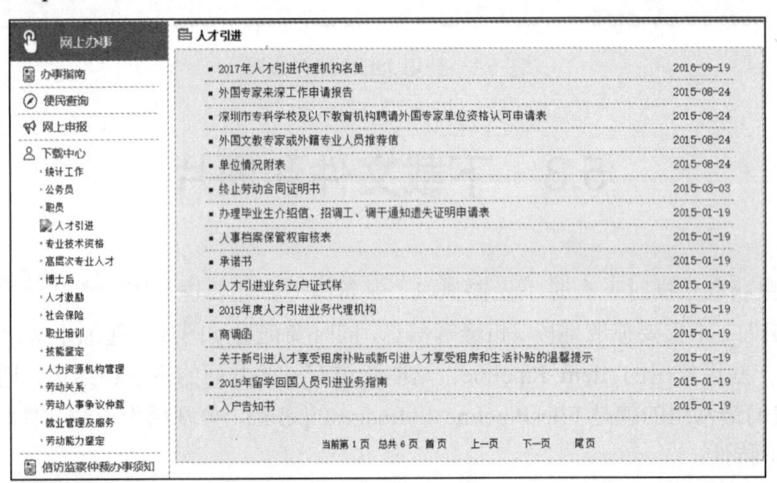

图 5.2 文件下载

（1）创建项目：

```
scrapy startproject filedownload
```

（2）创建爬虫：

```
scrapy genspider getfile hrss.sz.gov.cn
```

（3）在 items.py 中，要为 Item 添加两个字段，分别是文件 URL 字段和文件下载结果信息字段：

```
01 # -*- coding: utf-8 -*-
02
```

```
03 # Define here the models for your scraped items
04 #
05 # See documentation in:
06 # https://doc.scrapy.org/en/latest/topics/items.html
07
08 import scrapy
09
10
11 class DownloadfileItem(scrapy.Item):
12     # define the fields for your item here like:
13     # 文件名称
14     file_name = scrapy.Field()
15     # 发布时间
16     release_date = scrapy.Field()
17     # 文件URL
18     file_urls = scrapy.Field()
19     # 文件结果信息
20     files = scrapy.Field()
```

（4）在settings.py的ITEM_PIPELINE中添加FilesPipeline：

```
01 # Configure item pipelines
02 # See https://doc.scrapy.org/en/latest/topics/item-pipeline.html
03 ITEM_PIPELINES = {
04     'scrapy.pipelines.files.FilesPipeline':1,
05 }
```

（5）在settings.py中设置文件下载的路径、文件URL对应的Item字段、文件结果信息对应的Item字段：

```
FILES_STORE = 'D:\\Scrapy\\downloadfiles'
FILES_URLS_FIELD = 'file_urls'
FILES_RESULT_FIELD = 'files'
```

（6）设置文件存储之后，编写爬虫，spiders/getfile.py内容如下：

```
01 # -*- coding: utf-8 -*-
02 import scrapy
03 from downloadfile.items import DownloadfileItem
04
05
06 class GetfileSpider(scrapy.Spider):
07     name = 'getfile'
08     allowed_domains = ['szhrss.gov.cn']
09     start_urls = ['http://hrss.sz.gov.cn/wsbs/xzzx/rcyj/']
10
11     def parse(self, response):
12         files_list = response.css('.conRight_text_ul1 li')
```

```
13        for file in files_list:
14            item = DownloadfileItem()
15            item['file_name'] = file.css('a::text').extract_first()
16            item['release_date'] = file.css('span::text').extract_first()
17            # 由于获取到的 url 类似"./201501/P020170328745500534334.doc"
18            # 因此需要手动调整为完成的 url 格式
19            url = file.css('a::attr(href)').extract_first()
20            # file_urls 必须是 list 形式
21            item['file_urls'] = [response.url + url[1:]]
22            yield item
23
```

（7）运行爬虫，部分结果如图 5.3 所示。

```
2019-03-06 23:27:42 [scrapy.core.scraper] DEBUG: Scraped from <200 http://hrss.sz.gov.cn/wsbs/xzzx/rcyj/>
{'file_name': '终止劳动合同证明书',
 'file_urls': ['http://hrss.sz.gov.cn/wsbs/xzzx/rcyj//201207/P020120720516613347967.doc'],
 'files': [{'checksum': '71bcabfb80bba3ade49494f1e8c61c37',
            'path': 'full/终止劳动合同证明书.doc',
            'url': 'http://hrss.sz.gov.cn/wsbs/xzzx/rcyj//201207/P020120720516613347967.doc'}],
 'release_date': '2015-03-03'}
2019-03-06 23:27:42 [scrapy.core.scraper] DEBUG: Scraped from <200 http://hrss.sz.gov.cn/wsbs/xzzx/rcyj/>
{'file_name': '外国文教专家或外籍专业人员推荐信',
 'file_urls': ['http://hrss.sz.gov.cn/wsbs/xzzx/rcyj//201508/P020150824598848082535.docx'],
 'files': [{'checksum': '35fc98c8625d9e7fd7e49ffd3bf95c9c',
            'path': 'full/外国文教专家或外籍专业人员推荐信.docx',
            'url': 'http://hrss.sz.gov.cn/wsbs/xzzx/rcyj//201508/P020150824598848082535.docx'}],
 'release_date': '2015-08-24'}
2019-03-06 23:27:42 [scrapy.core.scraper] DEBUG: Scraped from <200 http://hrss.sz.gov.cn/wsbs/xzzx/rcyj/>
{'file_name': '办理毕业生介绍信、招调工、调干通知遗失证明申请表',
 'file_urls': ['http://hrss.sz.gov.cn/wsbs/xzzx/rcyj//201501/P020150119522234639888.doc'],
 'files': [{'checksum': 'd7cf89db052f803c73ed65f634d1ff39',
            'path': 'full/办理毕业生介绍信、招调工、调干通知遗失证明申请表.doc',
            'url': 'http://hrss.sz.gov.cn/wsbs/xzzx/rcyj//201501/P020150119522234639888.doc'}],
 'release_date': '2015-01-19'}
2019-03-06 23:27:42 [scrapy.core.scraper] DEBUG: Scraped from <200 http://hrss.sz.gov.cn/wsbs/xzzx/rcyj/>
{'file_name': '外国专家来深工作申请报告',
 'file_urls': ['http://hrss.sz.gov.cn/wsbs/xzzx/rcyj//201508/P020150824599927418956.doc'],
 'files': [{'checksum': '9d0c90ba3b1a2e76f046e13be12e79c0',
```

图 5.3　文件下载运行结果

（8）在 D:\Scrapy\downloadfiles 文件夹中会自动创建 full 文件夹，其中下载文件如图 5.4 所示，文件名是根据文件 URL 自动生成的 SHA1 哈希值。

图 5.4　文件下载保存结果

说明：SHA1 是安全哈希算法（Secure Hash Algorithm），适用于数字签名算法。

（9）可以看到默认存储的文件名无法分辨出下载的是什么文件，这时我们就需要自定义文件下载管道，进行一些自定义的设置。修改 pipelines.py，添加自定义文件管道，内容如下：

```
01 from scrapy.pipelines.files import FilesPipeline
02 from scrapy import Request
03
04 class DownloadfilePipeline(FilesPipeline):
05
06     # 修改 file_path 方法，使用提取的文件名保存文件
07     def file_path(self, request, response=None, info=None):
08         # 获取到 Request 中的 item
09         item = request.meta['item']
10         # 文件 URL 路径的最后部分是文件格式
11         file_type = request.url.split('.')[-1]
12         # 修改使用 item 中保存的文件名作为下载文件的文件名，文件格式使用提取到的
13         #格式
14         file_name = u'full/{0}.{1}'.format(item['file_name'],file_type)
15         return file_name
16
17     def get_media_requests(self, item, info):
18         for file_url in item['file_urls']:
19             # 为 request 带上 meta 参数，把 item 传递过去
20             yield Request(file_url,meta={'item':item})
```

（10）重新运行爬虫，下载到的文件如图 5.5 所示。

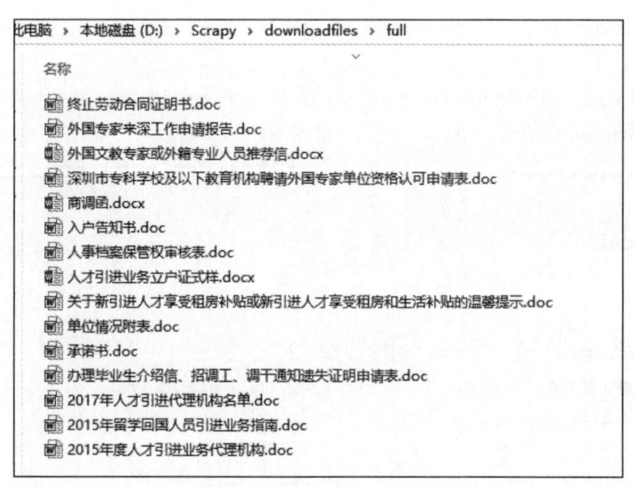

图 5.5　修改下载文件名

5.3.2　图片管道

图片管道与文件管道的工作流程一致，不同的是资源 URL 存储由 file_urls 变更为 image_urls，结果存储字段由 files 变为 images，具体流程如下：

（1）在爬虫中，抓取到 Item 并把期望的 URL 放入 image_urls 内。

（2）从爬虫内返回的 Item 进入 Item Pipeline 内。

（3）当Item进入FilesPipeline时，image_urls内的URL将被Scrapy内置的调度器和下载器安排下载，这就是说，调度器和下载器中间件是可以重复使用的。如果是更高的优先级，那么这些URL会在抓取其他页面之前被处理。在这个管道内的Item会保持锁定状态，直到文件下载完成或因其他原因下载失败。

当文件下载完成之后，下载的结果将会填充到images字段中。这个字段是由dict类型数据组成的列表，包含下载路径、源地址（从image_urls中获取）、图片校验码（checksum）。images字段中文件的顺序与image_urls文件URL顺序保持一致。如果下载失败，就会记录错误信息，文件并不会出现在images字段中。

【示例 5-4】通过抓取起点小说网完本小说（https://www.qidian.com/finish）的简介与封面图片来做实例演示

（1）创建项目：

```
>>>scrapy startproject downloadimage
```

（2）创建爬虫：

```
>>>cd downloadimage
>>>scrapy genspider getimage qidian.com
```

（3）创建项目之后，在items.py文件中添加图片下载对应的字段：

```
01 # -*- coding: utf-8 -*-
02
03 # Define here the models for your scraped items
04 # See documentation in:
05 # https://doc.scrapy.org/en/latest/topics/items.html
06
07 import scrapy
08
09
10 class ImgpipelineItem(scrapy.Item):
11     # define the fields for your item here like:
12     # name = scrapy.Field()
13     # 小说名称
14     title = scrapy.Field()
15     # 小说作者
16     author = scrapy.Field()
17     # 小说类型
18     type = scrapy.Field()
19     # 图片URL
20     image_urls = scrapy.Field()
21     # 图片结果信息
22     images = scrapy.Field()
```

（4）在 settings.py 的 ITEM_PIPELINE 中添加 ImagesPipeline 及自定义的管道信息：

```
# Configure item pipelines
# See https://doc.scrapy.org/en/latest/topics/item-pipeline.html
ITEM_PIPELINES = {
   'imgpipeline.pipelines.ImgpipelinePipeline': 300,
   'scrapy.pipeline.images.ImagesPipeline':1
}
```

（5）在 settings.py 中设置文件下载的路径、图片 URL 对应的 Item 字段、图片结果信息对应的 Item 字段，另外还可以设置图片的缩略图：

```
IMAGES_STORE = 'D:\\scrapy\\imgdownload'
IMAGES_URLS_FIELD = 'image_urls'
IMAGES_RESULT_FIELD = 'images'
IMAGES_THUMBS = {
    'small' : (80,80),
    'big' : (300,300)
    }
```

（6）编写自定义管道信息，用于保存小说信息至 JSON 文件，pipelines.py 内容如下：

```
01 # -*- coding: utf-8 -*-
02
03 # Define your item pipelines here
04 #
05 # Don't forget to add your pipeline to the ITEM_PIPELINES setting
06 # See: https://doc.scrapy.org/en/latest/topics/item-pipeline.html
07 import json
08
09 class DownloadimagePipeline(object):
10     # 将小说信息保存为json文件
11     def open_spider(self,spider):
12         self.file = open('qidian.json','w')
13
14     def close_spider(self,spider):
15         self.file.close()
16
17     def process_item(self, item, spider):
18         # 写入文件
19         line = json.dumps(dict(item), ensure_ascii=False) + "\n"
20         self.file.write(line)
21         return item
22
```

（7）编写爬虫文件，抓取小说信息，getimg.py 内容如下：

```
01 # -*- coding: utf-8 -*-
02 import scrapy
03 from imgpipeline.items import ImgpipelineItem
04
05
06 class GetimgSpider(scrapy.Spider):
07     name = 'getimg'
08     allowed_domains = ['qidian.com']
09     start_urls = ['https://www.qidian.com/finish']
10
11     def parse(self, response):
12         for novel in response.css(".all-img-list > li"):
13             item = ImgpipelineItem()
14             item['title'] = novel.xpath('.//h4/a/text()').extract_first()
15             item['author'] = novel.css('.name::text').extract_first()
16             item['type'] = novel.css('em + a::text').extract_first()
17             item['image_urls'] = ['https:' + novel.xpath('.//img/@src').
                                                    extract_first()]
18             yield item
```

（8）运行爬虫，会生成两个文件夹：full 和 thumbs。thumbs 中有 big 及 small 文件夹，保存的图片如图 5.6 所示。

图 5.6 图片下载结果

（9）与文件下载一样，默认图片名没有可读性，读者可仿照上一节中的自定义文件管道 pipeline.py 来自定义图片下载管道，修改下载图片名。

5.4 数据库存储 MySQL

5.3 节中讲到的信息存储都是存储到本地文件中，比如 JSON、TXT 等。实际运用中，保存到数据库中显然是一个更好的选择，数据安全、便于维护。本节介绍 MySQL 数据库的基本用法，以及如何使用 Python 对其进行相应的管理操作。

MySQL 原本是一个开放源代码的关系数据库管理系统（DBMS），原开发者为瑞典的 MySQL AB 公司，经过一些收购，现在 MySQL 成为 Oracle 旗下的产品。MySQL 由于性能高、成本低、可靠性好，已经成为最流行的开源数据库，被广泛地应用在互联网上的中小型网站中，是最流行的关系型数据库管理系统。本节讲解 MySQL 的一些基本操作。

5.4.1 在 Ubuntu 上安装 MySQL

在 Ubuntu 上安装 MySQL 的步骤如下：

步骤 01 先使用 wget 下载 MySQL 存储库软件包：

```
wget -c https://dev.mysql.com/get/mysql-apt-config_0.8.10-1_all.deb
```

步骤 02 进入文件下载的目录，然后使用以下 dpkg 命令安装下载好的 MySQL 存储库软件包：

```
sudo dpkg -i mysql-apt-config_0.8.10-1_all.deb
```

在软件包安装过程中，系统会提示选择 MySQL 服务器版本和其他组件，例如群集、共享客户端库或配置要安装 MySQL 的工作台。默认 MySQL 服务器版本 mysql-8.0 的源将被自动选中，我们只需最终确定就可以完成发行包的配置和安装，如图 5.7 所示。

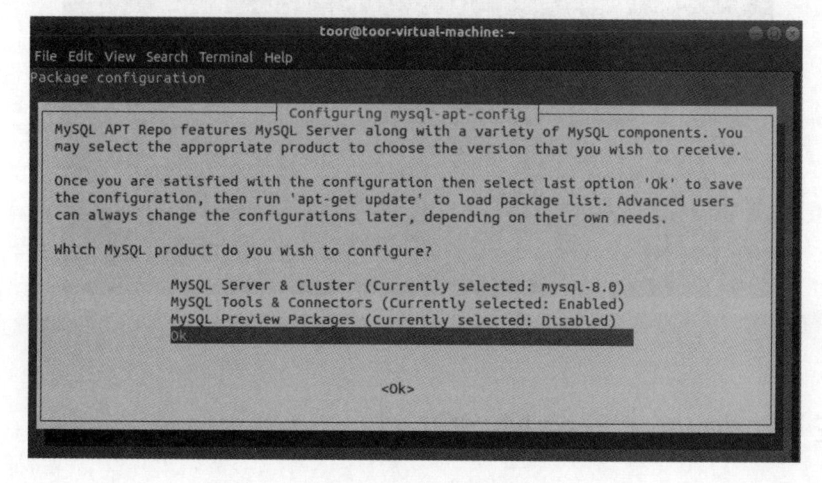

图 5.7 MySQL 版本选择

步骤 03 在 Ubuntu 18.04 中安装 MySQL 8 服务器。

先更新最新的软件包信息：

```
sudo apt update
```

然后运行如下命令安装 MySQL 8 社区服务器、客户端和数据库公用文件：

```
sudo apt-get install mysql-server
```

安装过程中将会要求为 MySQL 8 服务器的 root 设置用户输入密码，在输入和再次验证后按回车键继续。注意，输入完密码之后会进入选择默认密码加密方式的页面，如图 5.8 所示，此时要选择 Use legacy Authentication Method (Retain MySql 5.x Compatibility)选项，这是由于 Ubuntu18.04 的终端不支持第一种加密方式，如果选择第一种，就会登录不上去。

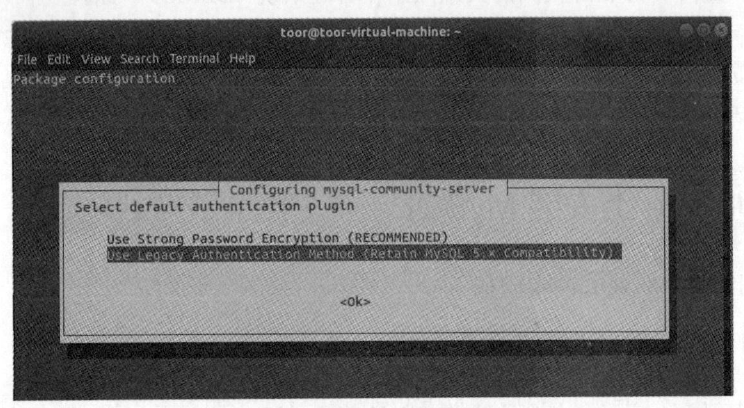

图 5.8 密码加密方式选择

步骤 04 安装完成之后，使用登录命令 mysql –uroot -p 登录数据库，如图 5.9 所示。

```
toor@toor-virtual-machine: ~
File Edit View Search Terminal Help
Setting up mysql-community-server (8.0.12-1ubuntu18.04) ...
update-alternatives: using /etc/mysql/mysql.cnf to provide /etc/mysql/my.cnf (my.cnf) in au
to mode
Created symlink /etc/systemd/system/multi-user.target.wants/mysql.service → /lib/systemd/sy
stem/mysql.service.
Setting up mysql-server (8.0.12-1ubuntu18.04) ...
Processing triggers for libc-bin (2.27-3ubuntu1) ...
Processing triggers for ureadahead (0.100.0-20) ...
Processing triggers for systemd (237-3ubuntu10) ...
toor@toor-virtual-machine:~$ mysql -uroot -p
Enter password:
Welcome to the MySQL monitor.  Commands end with ; or \g.
Your MySQL connection id is 8
Server version: 8.0.12 MySQL Community Server - GPL

Copyright (c) 2000, 2018, Oracle and/or its affiliates. All rights reserved.

Oracle is a registered trademark of Oracle Corporation and/or its
affiliates. Other names may be trademarks of their respective
owners.

Type 'help;' or '\h' for help. Type '\c' to clear the current input statement.

mysql>
```

图 5.9 登录数据库

5.4.2 在 Windows 上安装 MySQL

在 Windows 上安装 MySQL 的步骤如下：

步骤 01 进入社区版下载页面（https://dev.mysql.com/downloads/windows/installer/8.0.html），下载安装程序 mysql-installer-community-8.0.12.0.msi。

步骤 02 双击运行安装程序，安装时选择 Developer Default，单击 Next 按钮进行下一步，如图 5.10 所示。

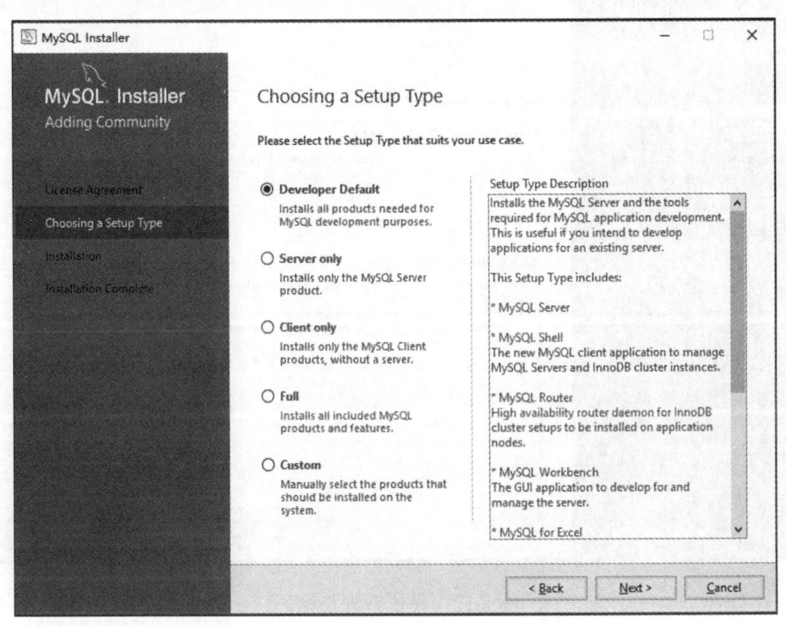

图 5.10 选择数据库类型

步骤 03 下一步检查依赖程序，如果缺少安装程序，那么可以单击 Execute 按钮安装依赖程序，单击 Next 按钮进行下一步，如图 5.11 所示。

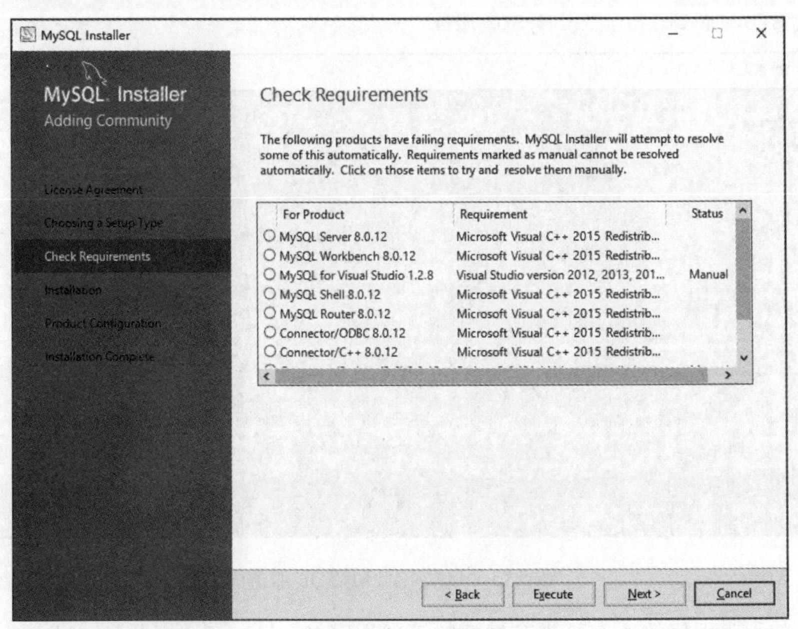

图 5.11 安装依赖程序

步骤04 之后进行安装，主程序安装完成后会进行账号配置，记住账号和密码，单击 Next 按钮进行下一步，如图 5.12 所示。

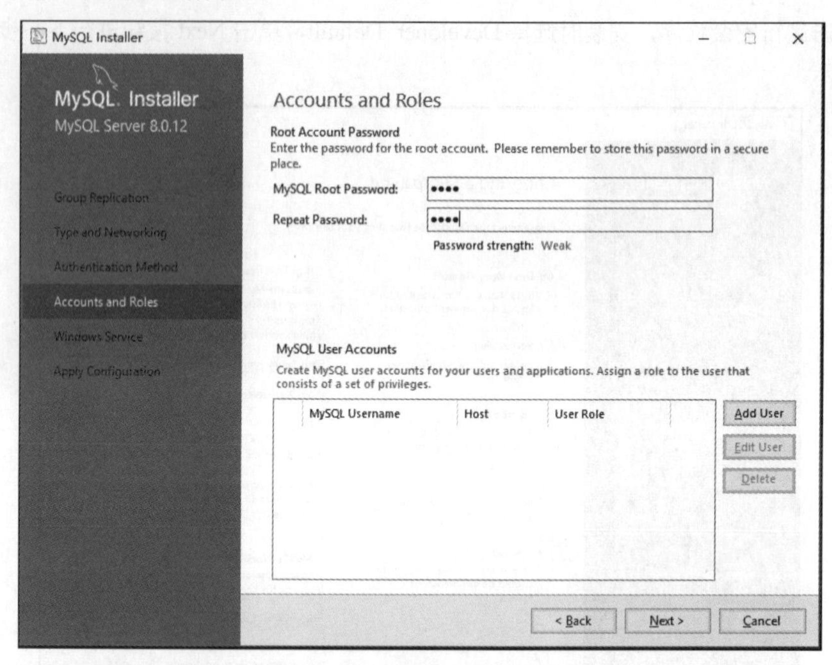

图 5.12 设置管理员密码

步骤05 之后只需默认单击 Next 按钮，即可安装完成，MySQL 进程会自动启动。以管理员身份打开命令行窗口，可以使用如下命令进行控制，如图 5.13 所示。

```
net start mysql80 启动 MySQL 服务
net stop mysql80 停止 MySQL 服务
```

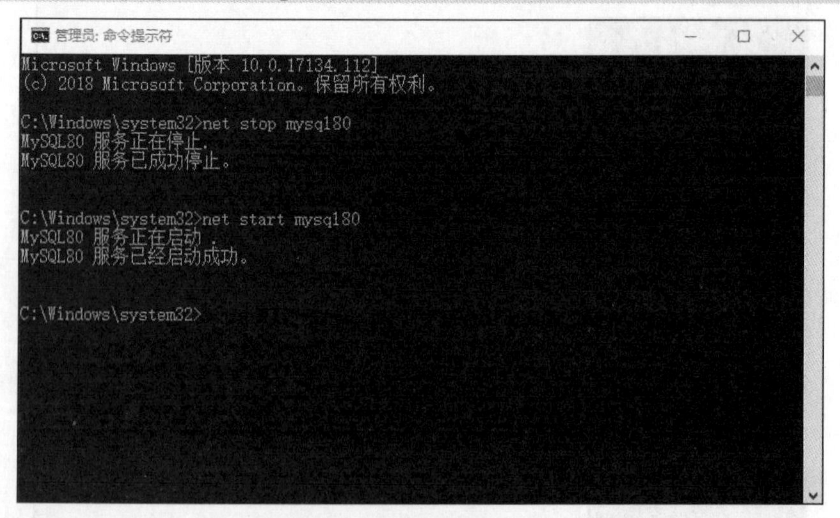

图 5.13 启动关闭 MySQL 服务

桌面端安装程序很好的一点是带有便捷的管理程序 Workbench。打开 Workbench，输入数据库密码即可连接管理数据库，使用起来很方便，如图 5.14 所示。

若想进入终端，可以在程序中选择 MySQL 8.0 Command Line Client，输入密码后即可进入，如图 5.15 所示。

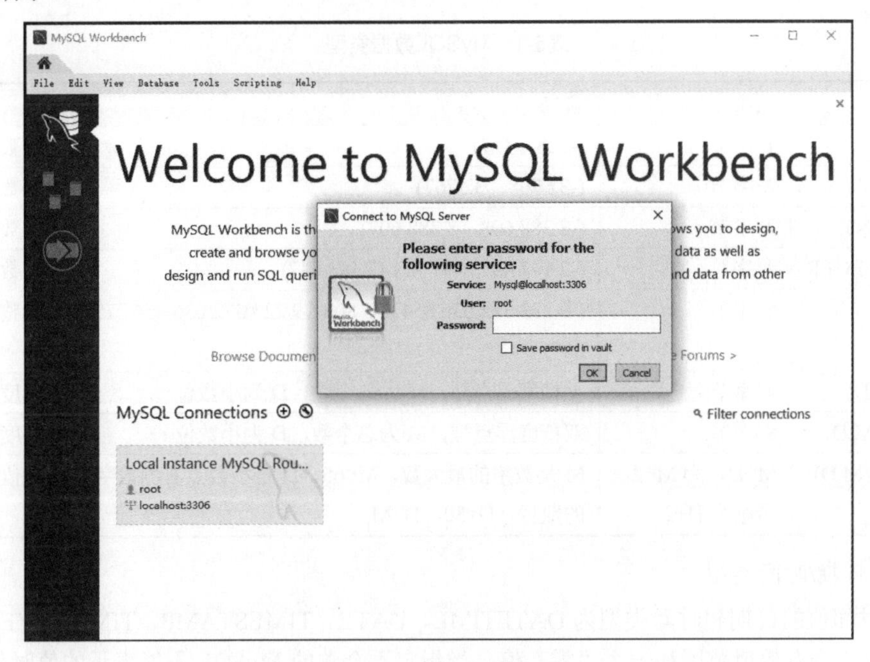

图 5.14 MySQL 使用 Workbench

```
MySQL 8.0 Command Line Client
Enter password: ****
Welcome to the MySQL monitor.  Commands end with ; or \g.
Your MySQL connection id is 12
Server version: 8.0.12 MySQL Community Server - GPL

Copyright (c) 2000, 2018, Oracle and/or its affiliates. All rights reserved.

Oracle is a registered trademark of Oracle Corporation and/or its
affiliates. Other names may be trademarks of their respective
owners.

Type 'help;' or '\h' for help. Type '\c' to clear the current input statement.

mysql>
```

图 5.15 MySQL 终端

5.4.3 MySQL 基础

安装 MySQL 之后，下面介绍 MySQL 的基础知识。

1. MySQL 数据类型

MySQL 支持多种类型，大致可以分为 3 类：数值、日期/时间和字符串（字符）类型，每种类型都有不同的范围与使用方法。

（1）数值类型

包括严格数值数据类型（TINYINT、MEDIUMINT、BIGINT、INTEGER、SMALLINT、DECIMAL

和 NUMERIC）和近似数值数据类型（FLOAT、REAL 和 DOUBLE PRECISION），汇总如表 5-1 所示。

表5-1　MySQL数据类型

类型	大小	含义	用途
TINYINT	1 字节	(-128，127)	小整数值
SMALLINT	2 字节	(-32 768，32 767)	大整数值
MEDIUMINT	3 字节	(-8 388 608，8 388 607)	大整数值
INT 或 INTEGER	4 字节	(-2 147 483 648，2 147 483 647)	大整数值
BIGINT	8 字节	(-9 233 372 036 854 775 808，9 223 372 036 854 775 807)	极大整数值
FLOAT(M,D)	4 字节	单精度浮点型，M 为总个数，D 为小数位	单精度浮点数值
DOUBLE(M,D)	8 字节	双精度浮点型，M 为总个数，D 为小数位	双精度浮点数值
DECIMAL(M,D)	M>D，为 M+2，否则为 D+2	M 是数字的最大数，M<65，D 是小数点右侧数字的数目，D<30，D<M	小数值

（2）日期/时间类型

表示时间值的日期和时间类型为 DATETIME、DATE、TIMESTAMP、TIME 和 YEAR。每个时间类型有一个有效值范围和一个“零”值，当指定不合法的 MySQL 不能表示的值时使用“零”值，汇总如表 5-2 所示。

表5-2　日期时间类型

类型	大小	格式	用途
DATE	3	YYYY-MM-DD	日期值
TIME	3	HH:MM:SS	时间值或持续时间
YEAR	1	YYYY	年份值
DATETIME	8	YYYY-MM-DD HH:MM:SS	混合日期和时间值
TIMESTAMP	4	YYYYMMDD HHMMSS	混合日期和时间值，时间戳，自动存储记录修改的时间

（3）字符串（字符）类型

字符串类型指 CHAR、VARCHAR、BINARY、VARBINARY、BLOB、TEXT、ENUM 和 SET，汇总如表 5-3 所示。

表5-3　字符串类型

类型	大小	用途
CHAR	0~255 字节	定长字符串
VARCHAR	0~65535 字节	变长字符串
TINYBLOB	0~255 字节	不超过 255 个字符的二进制字符串
TINYTEXT	0~255 字节	短文本字符串

（续表）

类型	大小	用途
BLOB	0~65 535 字节	二进制形式的长文本数据
TEXT	0~65 535 字节	长文本数据
MEDIUMBLOB	0~16 777 215 字节	二进制形式的中等长度文本数据
MEDIUMTEXT	0~16 777 215 字节	中等长度文本数据
LONGBLOB	0~4 294 967 295 字节	二进制形式的极大文本数据
LONGTEXT	0~4 294 967 295 字节	极大文本数据

2. MySQL 关键字

MySQL 含有一些和数据有关的关键字，如表 5-4 所示。

表5-4 MySQL关键字

关键字	含义
NULL	可包含 NULL 值
NOT NULL	不可包含 NULL 值
PRIMARY KEY	主键
DEFAULT	默认值
UNSIGNED	无符号类型
AUTO_INCREMENT	自动递增，适用于整数类型
CHARCTER SET	指定字符集，如 UTF-8、GBK 等

5.4.4 MySQL 基本操作

尽管使用 Workbench 很方便做 MySQL 的管理操作，但熟悉常用的命令仍是必不可少的。下面对数据库的常用操作命令进行介绍。

1. 创建与删除数据库

使用 create database database_name 指定数据库名称，创建一个数据库，创建成功后将返回一条成功信息：

```
mysql> create database test;
Query OK, 1 row affected (0.05 sec)
```

创建成功后，若想使用指定的数据库，则需要先选择后才能使用，使用 use database_name 命令进行选择：

```
mysql> use test;
Database changed
```

若想删除数据库，则使用 drop database database_name 命令进行删除：

```
mysql> drop database test;
Query OK, 0 rows affected (0.06 sec)
```

2. 创建与删除表

选择数据库之后，在此数据库中创建表，使用 create table table_name 命令：

```
mysql> create table book(
    -> id int unsigned not null auto_increment primary key,
    -> name char(50) not null,
    -> price float(5,2) not null
    -> );
Query OK, 0 rows affected (0.17 sec)
```

也可以将建表语句保存为 create_book.sql 文件，执行建表操作，如将 create_table.sql 文件保存在 C:\sqlfiles 中：

```
mysql> source c:\sqlfiles\create_book.sql
Query OK, 0 rows affected (0.09 sec)
```

使用 show tables 来查看数据库中的表：

```
Query OK, 0 rows affected (0.09 sec)
mysql> show tables;
+--------------------+
| Tables_in_test |
+--------------------+
| book          |
| book2         |
+--------------------+
2 rows in set (0.02 sec)
```

3. 数据基本操作

所有数据的操作归根到底就是增、删、改、查 4 种类型。我们已经在 test 数据库中创建了 book 表，下面对这张表进行数据操作。

插入数据使用 insert into table_name values(…)命令：

```
mysql> insert into book values(null,'python',45.56);
Query OK, 1 row affected (0.04 sec)
```

插入数据之后，使用 select column_name from table_name [where conditions]:

```
mysql> select * from book;
+----+--------+-------+
| id | name   | price |
+----+--------+-------+
|  1 | python | 45.56 |
+----+--------+-------+
1 row in set (0.00 sec)
mysql> select price from book;
```

```
+--------+
| price |
+--------+
| 45.56 |
+--------+
1 row in set (0.00 sec)
```

更新字段值使用 update table_name set column_name = value where conditions，如图 5.16 所示。

```
mysql> update book set price=48 where name='python';
Query OK, 1 row affected (0.09 sec)
Rows matched: 1  Changed: 1  Warnings: 0

mysql> _
```

图 5.16 更新数据

删除数据使用 delete from table_name where conditions，如图 5.17 所示。

```
mysql> delete from book where name='python';
Query OK, 1 row affected (0.01 sec)

mysql> _
```

图 5.17 删除数据

使用 alter table_name action 对表结构进行相关操作，如表 5-5 所示。

表5-5 alter命令

命令	说明
alter table book add publisher varchar(50);	在 ook 表中自动添加一列 publisher 到最后
alter table book drop publisher;	在 book 表中删除 publisher 列
alter table book change name title char(50)	在 book 表中将 name 列名更名为 title，数据类型更改为 char(50)
alter table book price double(6,2)	在 book 表中将 price 类数据类型更改为 double(6,2)
alter table book rename salebook	将 book 表更名为 salebook

如果想删除整张表，那么可以使用 drop table table_name 命令，如图 5.18 所示。

```
mysql> drop table book2;
Query OK, 0 rows affected (0.07 sec)

mysql>
```

图 5.18 删除表

5.4.5 Python 操作 MySQL

Python 中使用 pymysql 库来操作 MySQL，先使用 pip 进行安装，命令如下：

```
pip install pymysql
```

安装完成之后，导入模块：

```
>>> import pymysql
```

连接数据库使用 connnect 方法：

```
>>> db = pymysql.connect(host='localhost',port=3306,user='root',
password='root',database='test')
```

连接之后，使用 cursor()方法创建游标对象：

```
>>> cursor = db.cursor()
```

游标对象中常用的方法如下。

- execute()：执行 SQL 语句。
- executemany()：执行多条 SQL 语句。
- fetchall()：取出结果中所有数据。
- fetchmany()：取出结果中多条记录。
- fetchone()：取出结果中一条记录，并将游标指向下一条记录。
- close()：关闭游标对象。

前面已经连接上了 test 数据库，现在在这个数据库中对上面的方法进行讲解。

（1）创建 products 表，包含 id、name、price、amount：

```
>>> create_products = 'create table products ( id int not null auto_increment
primary key,name varchar(50) not null,price float(6,2) not null,amount int not
null)'
>>> cursor.execute(create_products)
0
```

（2）向 products 表中插入一条数据：

```
>>>
>>> cursor.execute('insert into products (name,price,amount) values
(%s,%s,%s)',('phone',1999.99,5))
1
```

向 products 表中插入多条数据：

```
>>> datas=[('coffee',12.3,100),('cup',9.9,60),('pen',25,180)]
>>> cursor.executemany('insert into products (name,price,amount) values
(%s,%s,%s)',datas)
3
```

执行完上面两条插入命令，查看数据库表数据，发现 products 表中数据仍为空，这是因为这两种插入数据的方法不会立即生效，需要对数据库对象进行提交操作：

```
>>>db.commit()
```

（3）查询数据：

```
>>> cursor.execute('select * from products')
4
>>> cursor.fetchone()
(1, 'phone', 1999.99, 5)
```

```
>>> cursor.execute('select * from products')
4
>>> cursor.fetchmany(2)
((1, 'phone', 1999.99, 5), (2, 'coffee', 12.3, 100))
>>> cursor.fetchall()
((1, 'phone', 1999.99, 5), (2, 'coffee', 12.3, 100), (3, 'cup', 9.9, 60), (4,
'pen', 25.0, 180))
>>> cursor.execute('select * from products')
4
>>> result = cursor.fetchall()
>>> for i in result:
...     print(i)
...
(1, 'phone', 1999.99, 5)
(2, 'coffee', 12.3, 100)
(3, 'cup', 9.9, 60)
(4, 'pen', 25.0, 180)
```

（4）修改、删除数据：

```
>>> cursor.execute('update products set price=%s where
name=%s',(9999,'phone'))
1
>>> cursor.execute('delete from products where name=%s',('coffee'))
1
```

修改和删除数据同样需要执行commit方法才能生效。使用完数据库之后，要进行关闭操作，释放资源：

```
>>>db.commit()
>>>db.close()
```

5.5 数据库存储MongoDB

抓取的数据也可以存储在MongoDB数据库中。MongoDB是由C++语言编写的，是一个基于分布式文件存储的、NoSQL类型的开源数据库系统。MongoDB将数据存储为一个文档，数据结构由键值（key=>value）对组成。MongoDB文档类似于JSON对象，字段值可以包含其他文档、数组及文档数组。MongoDB有很多优点，比如：

- 面向文档存储，操作起来比较简单和容易。
- 模式自由。
- 支持动态查询。
- 支持完全索引，包含内部对象。
- 文件存储格式为BSON（一种JSON的扩展）。

所以，MongoDB 非常适合在爬虫开发中做大规模数据的存储。

5.5.1 在 Ubuntu 上安装 MongoDB

在 Ubuntu 上安装和配置 MongoDB 的步骤如下。

步骤 01 导入 MongoDB 密钥：

```
    sudo apt-key adv --keyserver hkp://keyserver.ubuntu.com:80 --recv
9DA31620334BD75D9DCB49F368818C72E52529D4
```

步骤 02 创建一个源文件，用于更新下载：

```
    echo "deb [ arch=amd64 ] https://repo.mongodb.org/apt/ubuntu
bionic/mongodb-org/4.0 multiverse" | sudo tee
/etc/apt/sources.list.d/mongodb-org-4.0.list
```

步骤 03 更新源：

```
    sudo apt-get update
```

步骤 04 安装 MongoDB：

```
    sudo apt-get install -y mongodb-org
```

5.5.2 在 Windows 上安装 MongoDB

在 Windows 上安装 MongoDB 的步骤如下。

步骤 01 下载社区版安装程序（https://www.mongodb.com/download-center/community），如图 5.19 所示。

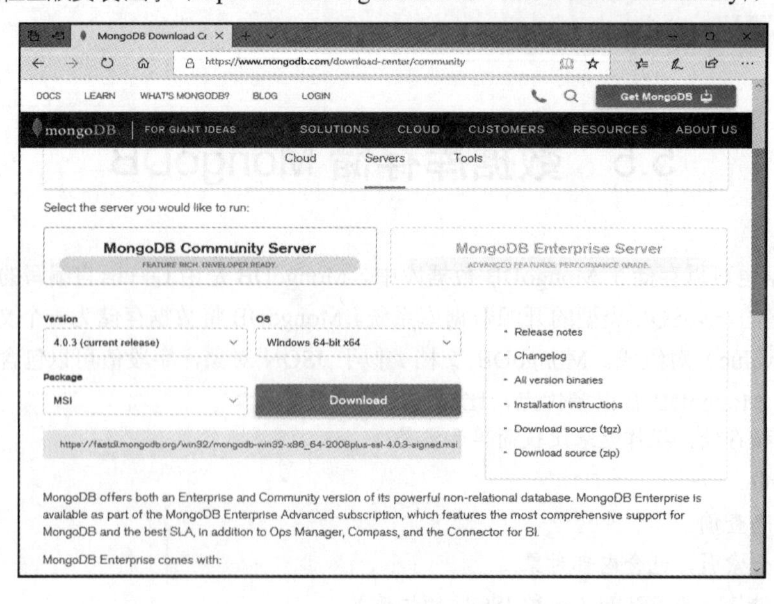

图 5.19 下载 MongoDB

步骤 02 下载完成之后进行安装，这里选择完全安装，如图 5.20 所示。

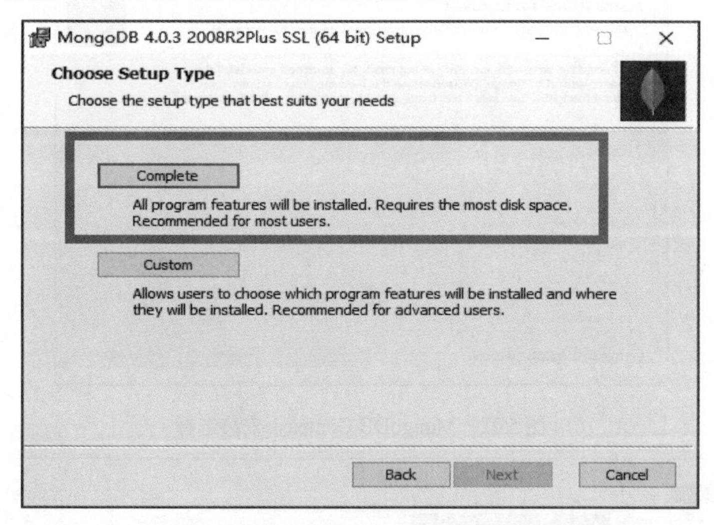

图 5.20 安装 MongoDB

下一步进入配置界面。这里选择将 MongoDB 配置为服务，Run service as Network Service user 表示可以远程连接数据库；Run service as a local or domain user 表示可以设置用户名密码验证信息进行登录。若有需要，则可以进行数据库路径与日志路径的配置，如图 5.21 所示。

MongoDB 4.0.3 2008R2Plus SSL (64 bit) Service Custo...
Service Configuration
Specify optional settings to configure MongoDB as a service.
Install MongoD as a Service
Run service as Network Service user
Run service as a local or domain user:
Account Domain: .
Account Name: MongoDB
Account Password:
Service Name: MongoDB
Data Directory: C:\Program Files\MongoDB\Server\4.0\data\
Log Directory: C:\Program Files\MongoDB\Server\4.0\log\
< Back　Next >　Cancel

图 5.21　MongoDB 安装配置

最后安装页面上会提示安装 MongoDB Compass，这是一个数据库图形管理页面，可以用来方便地查看数据库信息。如果网络环境不好，可以取消此复选框，不然安装会很慢，如图 5.22 所示。

步骤 03 MongoDB 的启动文件 mongo.exe 在安装目录的 bin 文件夹下，将 bin 文件夹目录添加至系统 path 路径下，可以快速启动数据库，如图 5.23 所示。

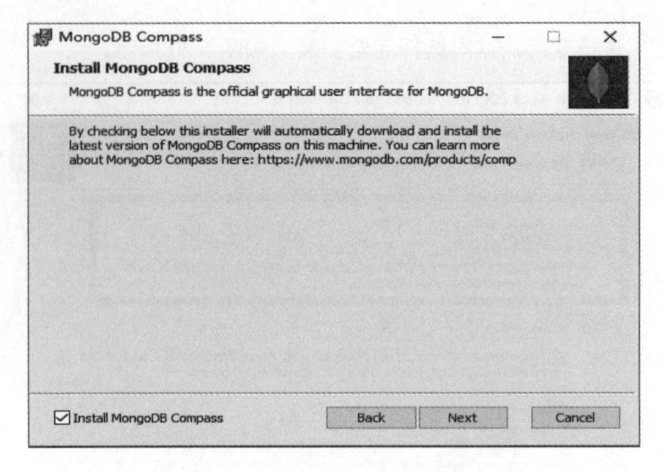

图 5.22　MongoDB Compass 安装选择

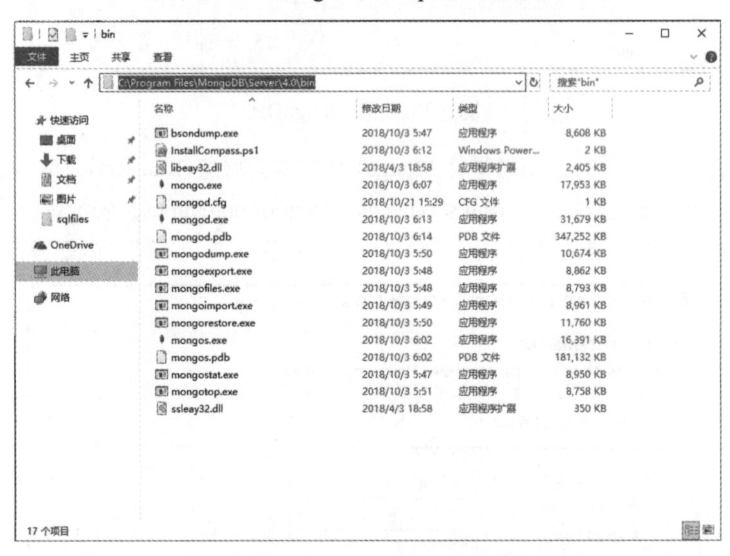

图 5.23　MongoDB 下的 bin 路径

使用 mongo 命令启动 MongoDB，如图 5.24 所示。

```
C:\Program Files\MongoDB\Server\4.0\bin>mongo
MongoDB shell version v4.0.3
connecting to: mongodb://127.0.0.1:27017
Implicit session: session { "id" : UUID("ef2a6a7d-c1cf-48e7-a238-bf932abd8976") }
MongoDB server version: 4.0.3
Server has startup warnings:
2018-10-21T15:30:24.619+0800 I CONTROL  [initandlisten]
2018-10-21T15:30:24.619+0800 I CONTROL  [initandlisten] ** WARNING: Access control is not enabled for the databa
se.
2018-10-21T15:30:24.619+0800 I CONTROL  [initandlisten] **          Read and write access to data and configurat
ion is unrestricted.
2018-10-21T15:30:24.619+0800 I CONTROL  [initandlisten]
---
Enable MongoDB's free cloud-based monitoring service, which will then receive and display
metrics about your deployment (disk utilization, CPU, operation statistics, etc).

The monitoring data will be available on a MongoDB website with a unique URL accessible to you
and anyone you share the URL with. MongoDB may use this information to make product
improvements and to suggest MongoDB products and deployment options to you.

To enable free monitoring, run the following command: db.enableFreeMonitoring()
To permanently disable this reminder, run the following command: db.disableFreeMonitoring()
---

>
```

图 5.24　启动 MongoDB

5.5.3 MongoDB 基础

MongoDB 属于 NoSQL 类型的数据库，与前面介绍的 MySQL 传统关系型数据库有很大差别。在 MongoDB 中，基本的概念是文档、集合、数据库，表 5-6 对两者做了对比。

表5-6 MongoDB与其他SQL的区别

SQL 概念术语	MongoDB 概念术语	说明
database	database	数据库
table	collection	数据库表/集合
row	document	数据库记录行/文档
column	field	数据字段/域
index	index	索引
primary key	primary key	主键，MongoDB 自动将_id 设置为主键

1. MongoDB 基本结构

（1）文档

MongoDB 的基本单元是文档，一个文档就是一条记录，文档以键值（key-value）方式存储数据，并且同一集合中的文档不需要有相同的字段，相同的字段也可以有不同的数据类型，这是 MongoDB 一个很突出的特点。一个简单的文档如下：

```
{"book":"python", "price":"45.99"}
```

需要注意的是：

- 文档的键值对是有序的，键值顺序不同，文档不同。下面两个文档是不同的：

```
{"book":"python", "price":"45.99"}
{"price":"45.99", "book":"python"}
```

- 文档区分类型和大小写，对比下面 4 个不同的文档：

```
{"book":"python", "price":45.99}
{"book":"python", "price":"45.99"}
{"PRICE":"45.99", "BOOK":"python"}
{"price":"45.99", "book":"python"}
```

- 文档的键不能重复，一个文档中一个键只能出现一次，这一点与 Python 中的字典数据类型一致。
- 文档的键除了少数情况外，可使用任意 UTF-8 字符：
  - 键不能含有\0（空字符），这个字符用来表示键的结尾。
  - “.”和“$”为系统保留，有特别的意义。
  - 以下画线“_”开头的键是保留的，建议不要使用。

（2）集合

集合就是 MongoDB 中的文档组，类似于 MySQL 中的数据表。集合存在于数据库中，集合没有固定的结构，这意味着我们对集合可以插入不同格式和类型的数据。比如，可以将以下不同数据结构的文档插入集合中：

```
{"book":"python", "price":"45.99"}
{"author":"Jim", "age":45, "works":[ "python", "java", "php"]}
```

集合也有一定的命名规则：

- 集合名不能是空字符串" "。
- 集合名不能含有\0 字符（空字符），这个字符表示集合名的结尾。
- 集合名不能以“system.”开头，这是为系统集合保留的前缀。
- 用户创建的集合名字不能含有保留字符。有些驱动程序的确支持在集合名里面包含，这是因为某些系统生成的集合中包含该字符。除非我们要访问这种系统创建的集合，否则千万不要在名字里出现$。

（3）数据库

一个 MongoDB 中可以建立多个数据库，其中默认数据库为“db”，该数据库存储在 data 目录中。MongoDB 的单个实例可以容纳多个独立的数据库，每一个都有自己的集合和权限，不同的数据库放置在不同的文件中。使用“show dbs”命令可以显示所有数据库，使用“db”命令查看当前使用的数据库。

2. MongoDB 数据类型

MongoDB 常用的数据类型如表 5-7 所示。

表5-7 MongoDB数据类型

数据类型	描述
String	字符串，存储数据常用的数据类型。在 MongoDB 中，UTF-8 编码的字符串才是合法的
Integer	整型数值，用于存储数值，根据采用的服务器可分为 32 位或 64 位
Boolean	布尔值
Double	双精度浮点值
Min/Max keys	将一个值与 BSON（二进制的 JSON）元素的最低值和最高值相对比
Arrays	用于将数组或列表或多个值存储为一个键
Timestamp	时间戳，记录文档修改或添加的具体时间
Object	用于内嵌文档
Null	用于创建空值
Symbol	符号，该数据类型基本上等同于字符串类型，但不同的是，它一般用于采用特殊符号类型的语言
Date	日期时间，用 UNIX 时间格式来存储当前日期或时间
Object ID	对象 ID，用于创建文档的 ID
Binary Data	二进制数据，用于存储二进制数据
Code	代码类型，用于在文档中存储 JavaScript 代码
Regular expression	正则表达式类型，用于存储正则表达式

5.5.4 MongoDB 基本操作

1. 数据库操作

使用 use DATABASE_NAME 创建数据库：

```
> use scrapy
switched to db scrapy
```

如果数据库不存在，就创建数据库，否则切换到指定的数据库。

使用 show dbs 查看所有的数据库：

```
>show dbs
admin        0.000GB
config       0.000GB
ithome       0.002GB
lianjia      0.000GB
local        0.000GB
segmentfault  0.004GB
```

上面我们创建的 scrapy 数据库没有显示出来，是因为 scrapy 数据库中没有数据。

```
> db.scrapy.insert({'python':'35.78'})
WriteResult({ "nInserted" : 1 })
> show dbs
admin        0.000GB
config       0.000GB
ithome       0.002GB
lianjia      0.000GB
local        0.000GB
scrapy       0.000GB
segmentfault  0.004GB
```

插入数据后，即可看到数据库 scrapy。

使用 db.dropDatabase()删除数据库：

```
> use scrapy
switched to db scrapy
> db.dropDatabase()
{ "dropped" : "scrapy", "ok" : 1 }
> show dbs
admin        0.000GB
config       0.000GB
ithome       0.002GB
lianjia      0.000GB
local        0.000GB
segmentfault  0.004GB
```

2. 集合操作

使用 db.createCollection(title,options)创建集合。其中，title 为集合名称，option 是可选的参数，可选参数如表 5-8 所示。

表5-8 MongoDB创建集合参数

字段	类型	描述
capped	布尔	可选参数，若为 true，则表示创建固定集合。固定集合是指有着固定大小的集合，当达到最大值时，它会自动覆盖最早的文档。当该值为 true 时，必须指定 size 参数
autoIndexId	布尔	可选参数，若为 true，则表示自动在_id 字段创建索引，默认为 false
size	数值	可选参数，为固定集合指定一个最大值（单位：字节）。如果 capped 为 true，就必须指定该字段
max	数值	可选参数，指定固定集合（当 capped=true 时）中包含文档的最大数量

在 book 数据库中创建名为 python 的集合：

```
> use book
switched to db book
> db.createCollection('python')
{ "ok" : 1 }
```

查看所有集合，使用 show collections 或 show tables：

```
> show collections
python
> show tables
python
```

使用 db.collection_name.drop()删除集合：

```
> db.python.drop()
true
> show collections
>
```

3. 文档操作

文档的数据结构和 JSON 基本一样。所有存储在集合中的数据都是 BSON 格式的。BSON 是类 JSON 的一种二进制形式的存储格式，是 Binary JSON 的简称。

（1）使用 db.collection_name.insert(document)在集合中插入文档：

```
> db.python.insert({'title':'Simple Python','author':'Jan Kin','price':45.89})
WriteResult({ "nInserted" : 1 })
```

使用 db.collection_name.insertMany([document1,document2,…])插入多条文档：

```
> db.python.insertMany([{'title':'Java','author':'Jams','price':55.00},
{'title':'精通 Python','author'
:'张三','price':36.78}])
{
        "acknowledged" : true,
        "insertedIds" : [
                ObjectId("5c77c2dce1af40396c67ae2f"),
                ObjectId("5c77c2dce1af40396c67ae30")
        ]
}
```

（2）使用 db.collection_name.find()查询文档：

```
> db.python.find()
{ "_id" : ObjectId("5c77bf7b082f4c56db42bae2"), "title" : "Simple Python",
"author" : "Jan Kin", "price" : 45.89 }
{ "_id" : ObjectId("5c77c2dce1af40396c67ae2f"), "title" : "Java", "author" :
"Jams", "price" : 55 }
{ "_id" : ObjectId("5c77c2dce1af40396c67ae30"), "title" : "精通 Python",
"author" : "张三", "price" : 36.78 }
```

可以使用.pretty()方法来使输出数据更易读：

```
> db.python.find().pretty()
{
        "_id" : ObjectId("5c77bf7b082f4c56db42bae2"),
        "title" : "Simple Python",
        "author" : "Jan Kin",
        "price" : 45.89
}
{
        "_id" : ObjectId("5c77c2dce1af40396c67ae2f"),
        "title" : "Java",
        "author" : "Jams",
        "price" : 55
}
{
        "_id" : ObjectId("5c77c2dce1af40396c67ae30"),
        "title" : "精通 Python",
        "author" : "张三",
        "price" : 36.78
}
```

在 find()方法中可以添加条件语句，用来更精确地查找结果，可用的条件语句如表 5-9 所示。

表5-9 MongoDB查询条件语句

操作	格式	示例	说明
等于	{<key>:<value>}	db.python.find({'author':'张三'})	从 python 集合中查找 author 为张三的文档
小于	{<key>:{$lt:<value>}}	db.python.find({'price':{$lt:50}})	从 python 集合中查找 price 小于 50 的文档
小于等于	{<key>:{$lte:<value>}}	db.python.find({'price':{$lte:45.89}})	从 python 集合中查找 price 小于等于 45.89 的文档
大于	{<key>:{$gt:<value>}}	db.python.find({'price':{$gt:50}})	从 python 集合中查找 price 大于 50 的文档
大于等于	{<key>:{$gte:<value>}}	db.python.find({'price':{$gte:45.89}})	从 python 集合中查找 price 大于等于 45.89 的文档
不等于	{<key>:{$ne:<value>}}	db.python.find({'price':{$ne:45.89}})	从 python 集合中查找 price 不等于 45.89 的文档

查询条件语句也可以组合使用，达到 AND 和 OR 的功能。将多个条件以逗号隔开，可以作为 AND 条件使用。比如查询集合中 author 为张三，并且 price 大于 40 的数据：

```
    > db.python.insert({'title':'Scrapy 入门','author':'张三','price':40.99})
    WriteResult({ "nInserted" : 1 })
    > db.python.find()
    { "_id" : ObjectId("5c77bf7b082f4c56db42bae2"), "title" : "Simple Python",
"author" : "Jan Kin", "price" : 45.89 }
    { "_id" : ObjectId("5c77c2dce1af40396c67ae2f"), "title" : "Java", "author" :
"Jams", "price" : 55 }
    { "_id" : ObjectId("5c77c2dce1af40396c67ae30"), "title" : "精通 Python",
"author" : "张三", "price" : 36.78 }
    { "_id" : ObjectId("5c77cac5e1af40396c67ae31"), "title" : "Scrapy 入门",
"author" : "张三", "price" : 40.99 }
    > db.python.find({'author':'张三','price':{$gt:40}}))
    { "_id" : ObjectId("5c77cac5e1af40396c67ae31"), "title" : "Scrapy 入门",
"author" : "张三", "price" : 40.99 }
```

使用关键字$or 可以执行 OR 条件功能。比如查询集合中 author 为张三或者 price 大于 45 的数据：

```
    > db.python.find({$or:[{'title':'张三'},{'price':{$gt:45}}]})
    { "_id" : ObjectId("5c77bf7b082f4c56db42bae2"), "title" : "Simple Python",
"author" : "Jan Kin", "price" : 45.89 }
    { "_id" : ObjectId("5c77c2dce1af40396c67ae2f"), "title" : "Java", "author" :
"Jams", "price" : 55 }
```

更复杂的条件语句可以结合 AND 与 OR 一起使用。下列语句查询 price 大于 40 的 author 为张三或者 title 为 Java 的数据：

```
    > db.python.find({'price':{$gt:40},$or:[{'author':'张三'},
{'title':'Java'}]})
    { "_id" : ObjectId("5c77c2dce1af40396c67ae2f"), "title" : "Java", "author" :
"Jams", "price" : 55 }
    { "_id" : ObjectId("5c77cac5e1af40396c67ae31"), "title" : "Scrapy 入门",
"author" : "张三", "price" : 40.99 }
```

（3）使用 update()更新文档，命令如下：

```
    db.collection.update(
       <query>,
       <update>,
       {
         upsert: <boolean>,
         multi: <boolean>,
         writeConcern: <document>
       }
    )
```

参数说明：

- query：update 的查询条件，类似于 where 的子句。
- update：update 的对象和一些更新的操作符，类似于 set 后面的语句。
- upsert：可选，如果不存在 update 的记录，是否插入新文档。True 为插入，默认是 False，不插入。
- multi：可选，默认是 False，只更新找到的第一条记录。如果这个参数为 True，就把按条件查出来的多条记录全部更新。
- writeConcern：可选，抛出异常的级别。

我们将 author 为“张三”的数据修改为 author 等于“张起灵”：

```
    > db.python.update({'author':'张三'},{$set:{'author':'张起灵'}},
{multi:true})
    WriteResult({ "nMatched" : 2, "nUpserted" : 0, "nModified" : 2 })
    > db.python.find().pretty()
    {
           "_id" : ObjectId("5c77bf7b082f4c56db42bae2"),
           "title" : "Simple Python",
           "author" : "Jan Kin",
           "price" : 45.89
    }
    {
           "_id" : ObjectId("5c77c2dce1af40396c67ae2f"),
           "title" : "Java",
           "author" : "Jams",
```

```
        "price" : 55
}
{
        "_id" : ObjectId("5c77c2dce1af40396c67ae30"),
        "title" : "精通 Python",
        "author" : "张起灵",
        "price" : 36.78
}
{
        "_id" : ObjectId("5c77cac5e1af40396c67ae31"),
        "title" : "Scrapy 入门",
        "author" : "张起灵",
        "price" : 40.99
}
```

（4）使用 remove()方法删除文档，命令格式如下：

```
db.collection.remove(
   <query>,
   {
     justOne: <boolean>,
     writeConcern: <document>
   }
)
```

参数说明：

- query：可选参数，删除文档的条件。
- justOne：可选参数，如果设为 True 或 1，就只删除一个文档。如果不设置该参数，或使用默认值 False，就删除所有匹配条件的文档。
- writeConcern：可选参数，抛出异常的级别。

删除 author 为张起灵的数据，只删除一条：

```
> db.python.remove({'author':'张起灵'},{justOne:1})
WriteResult({ "nRemoved" : 1 })
> db.python.find().pretty()
{
        "_id" : ObjectId("5c77bf7b082f4c56db42bae2"),
        "title" : "Simple Python",
        "author" : "Jan Kin",
        "price" : 45.89
}
{
        "_id" : ObjectId("5c77c2dce1af40396c67ae2f"),
```

```
        "title" : "Java",
        "author" : "Jams",
        "price" : 55
}
{
        "_id" : ObjectId("5c77cac5e1af40396c67ae31"),
        "title" : "Scrapy入门",
        "author" : "张起灵",
        "price" : 40.99
}
```

5.5.5 Python 操作 MongoDB

Python 操作 MongoDB 需要安装 pymongo 库，安装方法如下：

```
pip install pymongo
```

使用前导入 pymongo 库：

```
import pymongo
```

（1）连接 MongoDB

使用 pymongo 中的 MongoClient 模块连接 MongoDB，有 3 种方式：

- client = pymongo.MongoClient()
- client = pymongo.MongoClient(host,port)
- client = pymongo.MongoClient('mongodb://user:password@example.com:27017/test')

第 1 种是默认 MongoDB 的链接地址为 localhost，端口为 27017；第 2 种手动指定 host 和 port；第 3 种则传入一个完整的 URI 进行连接，指定用户名、密码、链接地址、端口与使用的数据库。

（2）获取数据库

连接上 MongoDB 之后，下一步就需要获取需要使用的数据库，有两种方式可用：

- db = client.book：使用属性访问。
- db = client['book']：使用字典形式访问。

（3）获取集合

与获取数据库一样，同样有访问属性与使用字典两种方式获取集合：

```
collection = db.python
collection = db['python']
```

（4）操作文档

操作文档的方法与前面演示的在 MongoDB Shell 中的操作方法基本一致，如查询依然使用 find()方法，并且查询条件同样适用，代码如下：

```
import pymongo

client = pymongo.MongoClient()

db = client['book']
# db = client.book
collection = db['python']
# collection = db.python

data1 = collection.find()
for d in data1:
    print(d)
print('-' * 15)
data2 = collection.find({'author':'张起灵'})
for d in data2:
    print(d)
```

运行结果：

```
{'_id': ObjectId('5c77bf7b082f4c56db42bae2'), 'name': 'Simple Python', 'author': 'Jan Kin', 'price': 45.89}
{'_id': ObjectId('5c77c2dce1af40396c67ae2f'), 'name': 'Java', 'author': 'Jams', 'price': 55.0}
{'_id': ObjectId('5c77cac5e1af40396c67ae31'), 'name': 'Scrapy 入门', 'author': '张起灵', 'price': 40.99}
---------------
{'_id': ObjectId('5c77cac5e1af40396c67ae31'), 'name': 'Scrapy 入门', 'author': '张起灵', 'price': 40.99}
```

5.6 实战：爬取链家二手房信息并保存到数据库

本节以爬取链家二手房信息数据为例来演示使用数据库保存数据、图片管道下载图片的方法，方便读者更好地理解与使用。本项目中编辑器为 PyCharm，数据库为 MongoDB，除了 MongoDB 官方提供的 MongoDB Compass 数据库查看工具外，PyCharm 中有一个很好用的 MongoDB 插件，可以方便地查看数据库数据。MongoDB 插件安装方法如下：

（1）进入 file-settings 界面，单击 Plugins 选项。

（2）在 Plugins 页面单击页面下方的 Browse repositories…按钮。

（3）在搜索框中搜索 Mongo Plugin，进行安装，如图 5.25 所示。

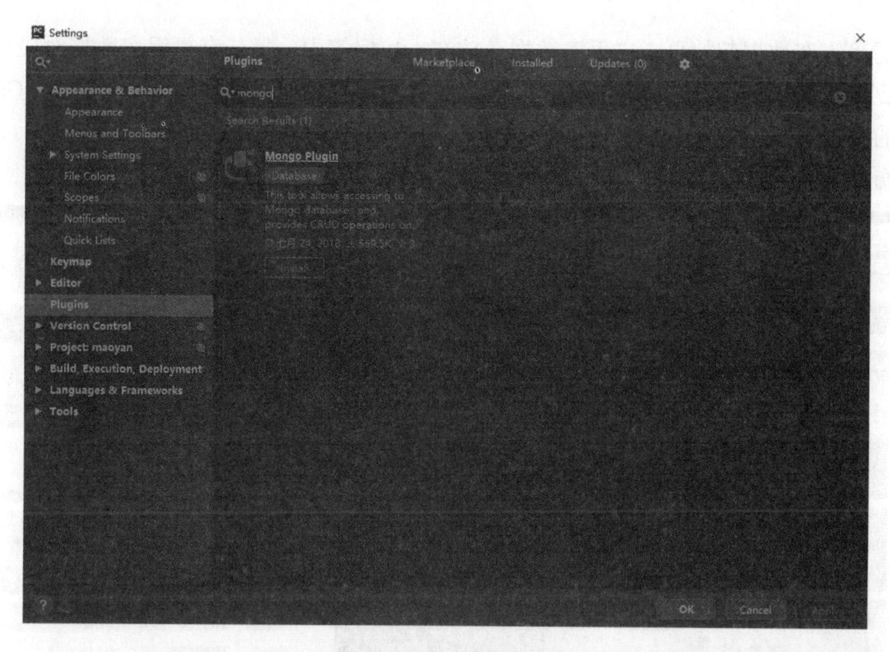

图 5.25 安装 Mongo Plugin

【示例 5-4】链家二手房房源数据抓取

先来分析起始数据页 https://bj.lianjia.com/ershoufang/的结构，如图 5.26 所示。

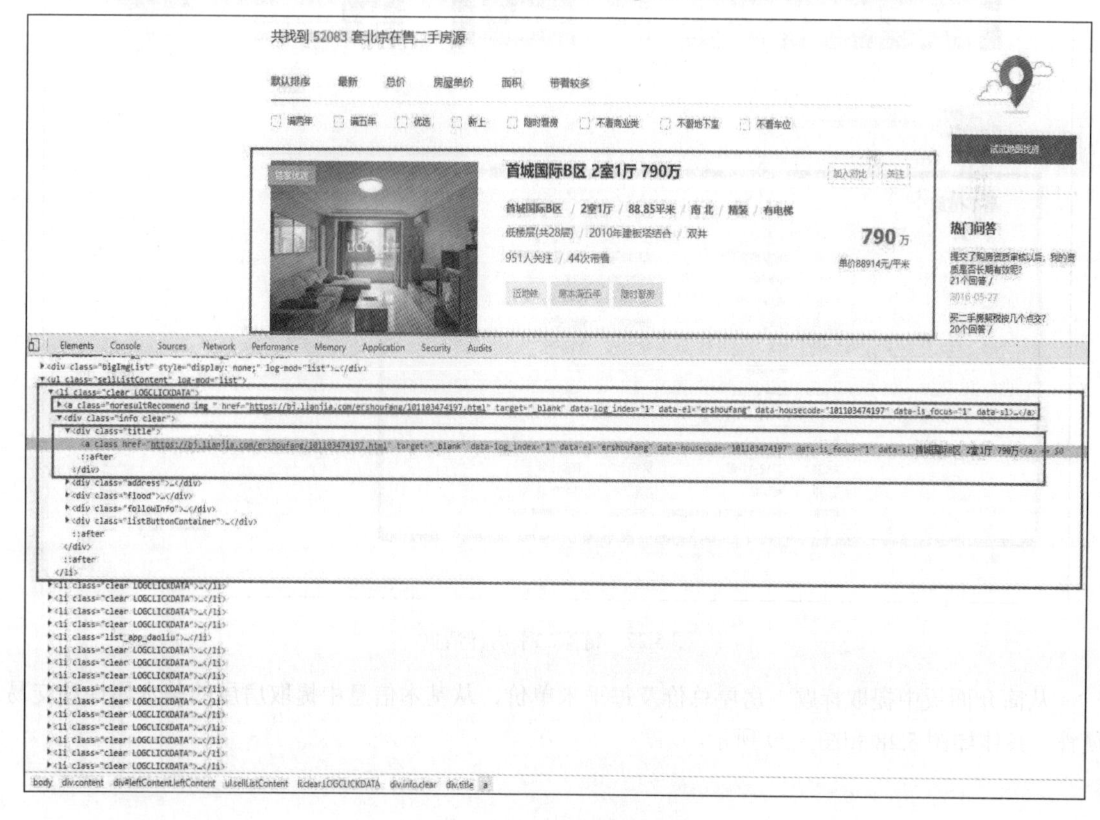

图 5.26 链家页面数据结构分析

可以看到，在数据起始页，二手房数据存放在<ul>标签中，每一套二手房数据对应一个<li>标签，<li>标签中的图片与标题的链接对应这个二手房的详细页面，因此我们可以使用 CrawlSpider 进行自动的跟进爬取。

再来分析详情页面，如图 5.27 所示。

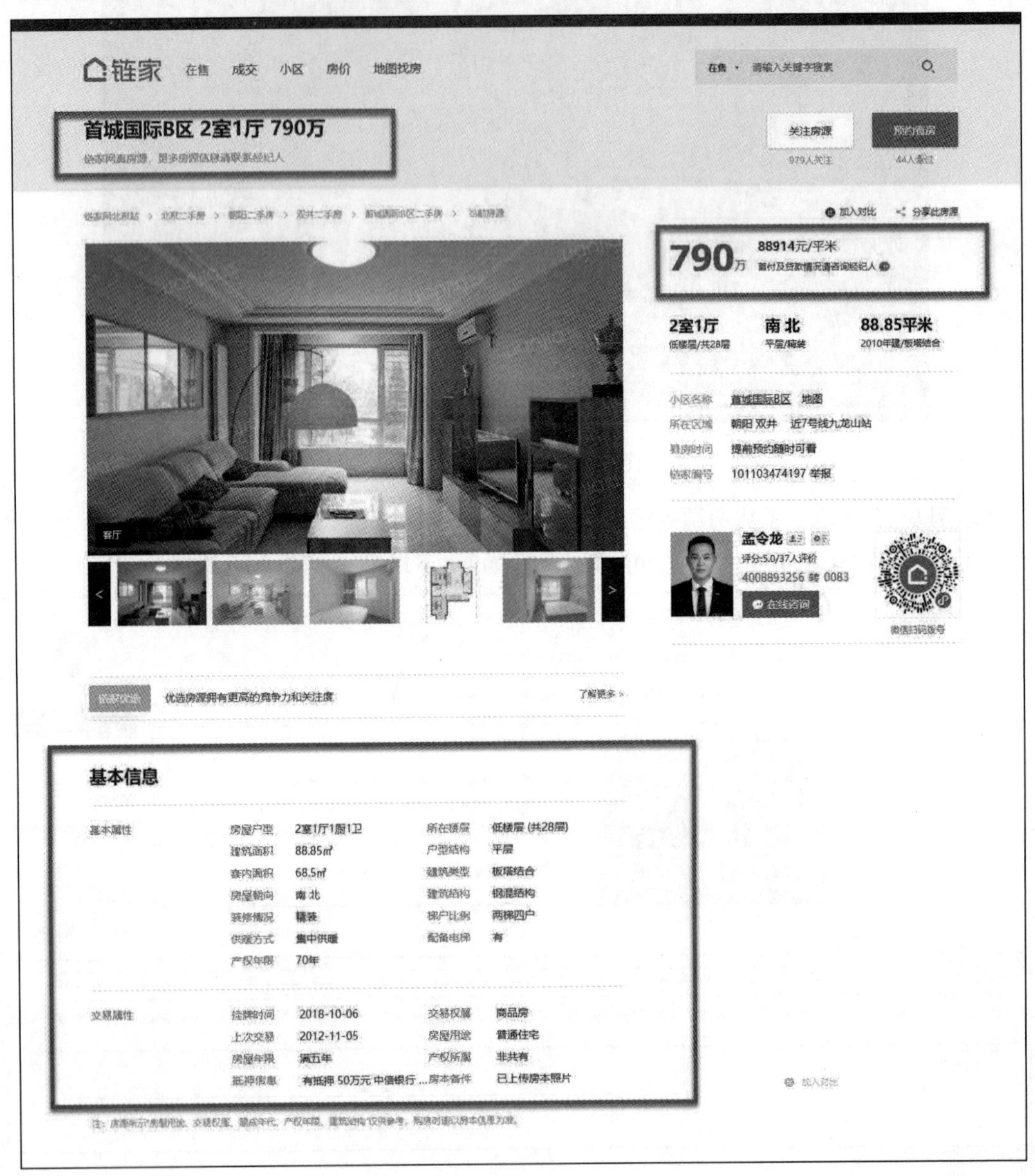

图 5.27 链家二手房详情页

从简介面板中提取标题、房屋总价及每平米单价，从基本信息中提取房屋的基本属性和交易属性，具体如图 5.28 和图 5.29 所示。

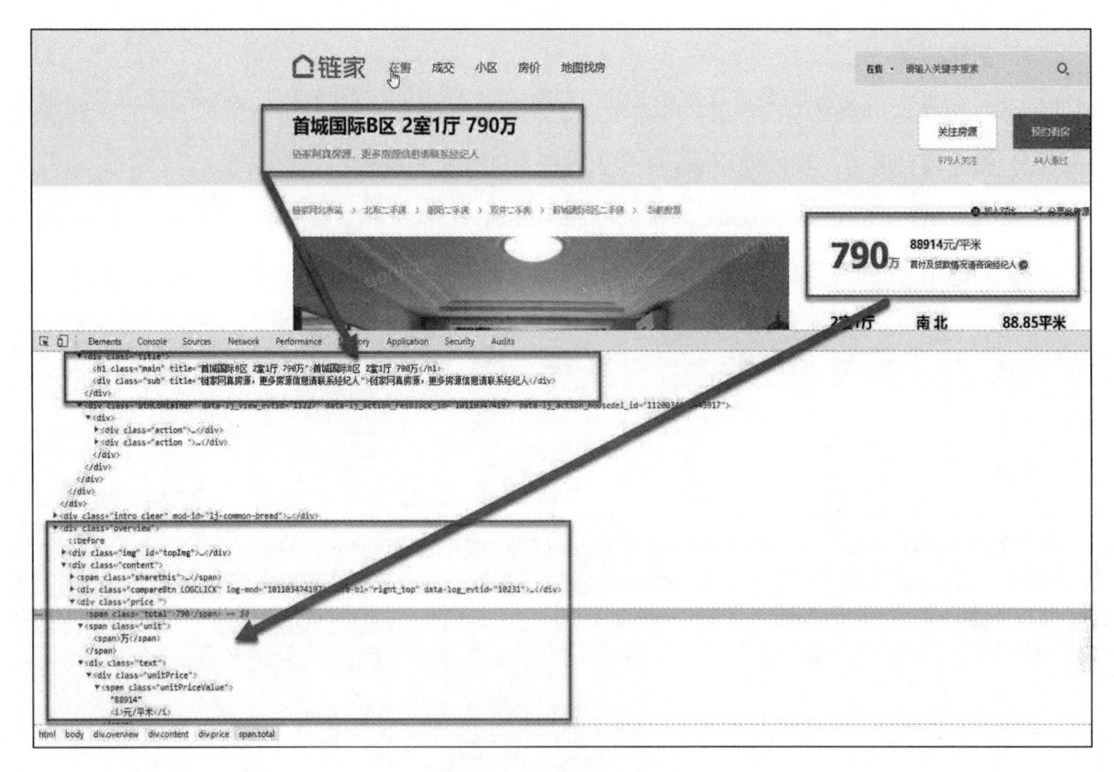

图 5.28　标题与价格定位

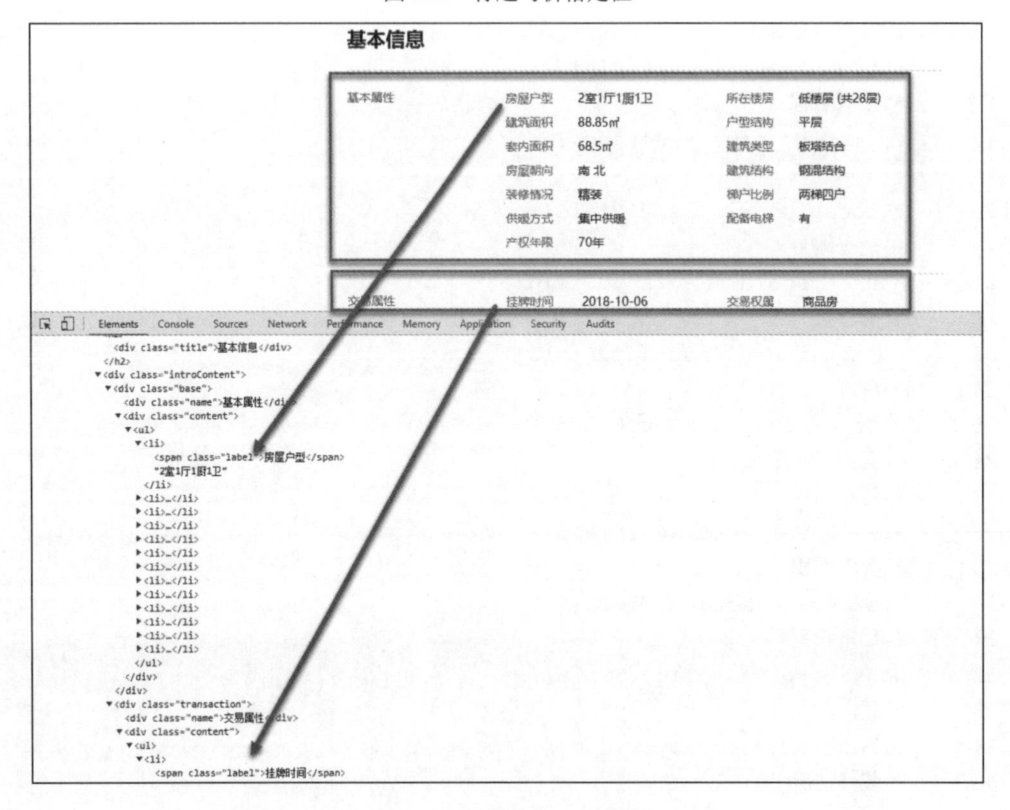

图 5.29　房屋属性元素定位

（1）创建项目：

```
scrapy startproject lianjiahouse
```

（2）创建项目之后，我们使用crawl爬虫模板来创建爬虫文件：

```
scrapy genspider -t crawl house lianjia.com
```

（3）创建项目之后，编写items.py文件，设置需要抓取的内容：

```
01 scrapy genspider -t crawl lianjiahouse lianjia.com
02 # -*- coding: utf-8 -*-
03
04 # Define here the models for your scraped items
05 #
06 # See documentation in:
07 # https://doc.scrapy.org/en/latest/topics/items.html
08
09 import scrapy
10
11
12 class LianjiaItem(scrapy.Item):
13     # define the fields for your item here like:
14     # name = scrapy.Field()
15     # 发布信息名称
16     house_name = scrapy.Field()
17     # 小区名称
18     community_name = scrapy.Field()
19     # 所在区域
20     # location = scrapy.Field()
21     # 链家编号
21     house_record = scrapy.Field()
22     # 总售价
23     total_amount = scrapy.Field()
24     # 单价
25     unit_price = scrapy.Field()
26     # 房屋基本信息
27     # 建筑面积
28     area_total = scrapy.Field()
29     # 套内面积
30     area_use = scrapy.Field()
31     # 厅室户型
32     house_type = scrapy.Field()
33     # 朝向
34     direction = scrapy.Field()
35     # 装修情况
36     sub_info = scrapy.Field()
```

```
37      # 供暖方式
38      heating_method = scrapy.Field()
39      # 产权
40      house_property = scrapy.Field()
41      # 楼层
42      floor = scrapy.Field()
43      # 总层高
44      total_floors = scrapy.Field()
45      # 电梯
46      is_left = scrapy.Field()
47      # 户梯比例
48      left_rate = scrapy.Field()
49      # 户型结构
50      structure = scrapy.Field()
51      # 房屋交易信息
52      # 挂牌时间
53      release_date = scrapy.Field()
54      # 上次交易时间
55      last_trade_time = scrapy.Field()
56      # 房屋使用年限
57      house_years = scrapy.Field()
58      # 房屋抵押信息
59      pawn = scrapy.Field()
60      # 交易权属
61      trade_property = scrapy.Field()
62      # 房屋用途
63      house_usage = scrapy.Field()
64      # 产权所有
65      property_own = scrapy.Field()
66      # 图片地址
67      images_urls = scrapy.Field()
68      # 保存图片
69      images = scrapy.Field()
```

（4）编写好 items.py 之后，然后编写 pipelines.py。这里主要编写两个 pipeline 方法：一个用于保存数据；另一个用于处理图片，代码如下：

```
01 # -*- coding: utf-8 -*-
02
03 # Define your item pipelines here
04 #
05 # Don't forget to add your pipeline to the ITEM_PIPELINES setting
06 # See: https://doc.scrapy.org/en/latest/topics/item-pipeline.html
07 import pymongo
08 from scrapy.pipelines.images import ImagesPipeline
09 from scrapy import Request
```

```
10
11 class LianjiaPipeline(object):
12     # 设置存储文档名称
13     collection_name = 'secondhandhouse'
14
15     def __init__(self, mongo_uri, mongo_db):
16         self.mongo_uri = mongo_uri
17         self.mongo_db = mongo_db
18
19     @classmethod
20     def from_crawler(cls, crawler):
21         return cls(
22             # 通过 crawler 获取 settings 文件，获取其中的 MongoDB 配置信息
23             mongo_uri=crawler.settings.get('MONGO_URI'),
24             mongo_db=crawler.settings.get('MONGO_DATABASE', 'lianjia')
25         )
26
27     def open_spider(self, spider):
28         # 当爬虫打开时连接 MongoDB 数据库
29         # 先连接 Server，再连接指定数据库
30         self.client = pymongo.MongoClient(self.mongo_uri)
31         self.db = self.client[self.mongo_db]
32
33     def close_spider(self, spider):
34         # 爬虫结束时关闭数据库连接
35         self.client.close()
36
37     def process_item(self, item, spider):
38         # 将 item 插入数据库
39         self.db[self.collection_name].insert(dict(item))
40         return item
41
42
43 class LianjiaImagePipeline(ImagesPipeline):
44     def get_media_requests(self, item, info):
45         for image_url in item['images_urls']:
46             # 将图片地址传入 Request，进行下载，同时将 item 参数添加到 Request 中
47             yield Request(image_url, meta={'item': item})
48
49     def file_path(self, request, response=None, info=None):
50         # 从 Request 中获取 item，以房屋标题作为文件夹名称
51         item = request.meta['item']
52         image_folder = item['house_name']
53         # 使用图片 URL 作为图片存储名称
54         image_guild = request.url.split('/')[-1]
```

```
55          # 图片保存，文件夹/图片
56          image_save = u'{0}/{1}'.format(image_folder, image_guild)
57          return image_save
```

（5）在settings.py中激活Pipeline，设置图片存储信息、MongoDB数据库信息。还有一点要注意，将REBOTSTXT_OBEY属性改为False，默认遵守robots协议的话是不能爬取信息的，后续章节将会详细讲解Scrapy默认设置：

```
# Obey robots.txt rules
ROBOTSTXT_OBEY = False

# Configure item pipelines
# See https://doc.scrapy.org/en/latest/topics/item-pipeline.html
ITEM_PIPELINES = {
   'lianjia.pipelines.LianjiaPipeline': 300,
   'lianjia.pipelines.LianjiaImagePipeline':400
}

# 图片存储配置
IMAGES_STORE = 'D:\\Scrapy\\chapter5\\5_04\\lianjia\\images'
IMAGES_URLS_FIELD = 'images_urls'
IMAGES_RESULT_FIELD = 'images'

# MongoDB 配置信息
MONGO_URI = 'localhost:27017'
MONGO_DATABASE = 'lianjia'
```

（6）编写爬虫文件，其中的Rule要追踪每条房屋信息的详情页面：

```
01 # -*- coding: utf-8 -*-
02 import scrapy
03 from scrapy.linkextractors import LinkExtractor
04 from scrapy.spiders import CrawlSpider, Rule
05 from lianjia.items import LianjiaItem
06 class SechandhouseSpider(CrawlSpider):
07     name = 'lianjiahouse'
08     allowed_domains = ['lianjia.com']
09     start_urls = ['https://bj.lianjia.com/ershoufang/']
10
11     rules = (
12         Rule(LinkExtractor(allow='/ershoufang/\d{12}.html'),
                                callback='parse_item'),
13     )
14
15     def parse_item(self, response):
16         i = LianjiaItem()
17         # 二手房名称
```

```
18        i['house_name'] = response.css
                    ('title::text').extract_first().replace(' ','')
19        # 所在小区
20        i['community_name'] = response.css('.communityName a::text').
                                                    extract_first()
21        # i['location'] = response.css()
22        # 链家编号
23        i['house_record'] = response.css
                        ('.houseRecord .info::text').extract_first()
24        # 总价
25        i['total_amount'] = response.css
                        ('.overview .total::text').extract_first()
26        # 房屋信息
27        # 单价
28        i['unit_price'] = response.css
                        ('.unitPriceValue::text').extract_first()
29        # 建筑总面积
30        i['area_total'] = response.xpath
                        ('//div[@class="base"]//ul/li[3]/text()')\
31            .re_first('\d+.\d+')
32        # 使用面积
33        i['area_use'] = response.xpath
                        ('//div[@class="base"]//ul/li[5]/text()')\
34            .re_first('\d+.\d+')
35        # 房屋类型
36        i['house_type'] = response.xpath
                        ('//div[@class="base"]//ul/li[1]/text()')\
37            .extract_first()
38        # 房屋朝向
39        i['direction'] = response.xpath
                        ('//div[@class="base"]//ul/li[7]/text()')\
40            .extract_first()
41        # 装修情况
42        i['sub_info'] = response.xpath
                        ('//div[@class="base"]//ul/li[9]/text()')\
43            .extract_first()
44        # 供暖方式
45        i['heating_method'] = response.xpath
                        ('//div[@class="base"]//ul/li[11]/text()')\
46            .extract_first()
47        # 产权
48        i['house_property'] = response.xpath
                        ('//div[@class="base"]//ul/li[13]/text()')\
49            .extract_first()
50        # 楼层
```

```
51          i['floor'] = response.xpath
                        ('//div[@class="base"]//ul/li[2]/text()')\
52              .extract_first()
53          # 总楼层
54          i['total_floors'] = response.xpath
                        ('//div[@class="base"]//ul/li[2]/text()')\
55              .re_first(r'\d+')
56          # 是否有电梯
57          i['is_left'] = response.xpath
                        ('//div[@class="base"]//ul/li[12]/text()')\
58              .extract_first()
59          # 户梯比例
60          i['left_rate'] = response.xpath
                        ('//div[@class="base"]//ul/li[10]/text()')\
61              .extract_first()
62          # 挂牌时间
63          i['release_date'] = response.xpath
                            ('//div[@class="transaction"]//ul/li[1]'
64                           '/span[2]/text()').extract_first()
65          # 最后交易时间
66          i['last_trade_time'] = response.xpath
                            ('//div[@class="transaction"]//ul/li[3]'
67                           '/span[2]/text()').extract_first()
68          # 房屋使用年限
69          i['house_years'] = response.xpath
                            ('//div[@class="transaction"]//ul/li[5]'
70                           '/span[2]/text()').extract_first()
71          # 房屋抵押信息，抵押信息中有空格及换行符，
72          # 先通过replace()将空格去掉，再通过strip()将换行符去掉
73          i['pawn'] = response.xpath
                    ('//div[@class="transaction"]//ul/li[7]/span[2]'
74                   '/text()').extract_first().replace(' ','').strip()
75          # 交易权属
76          i['trade_property'] = response.xpath
                        ('//div[@class="transaction"]//ul/li[2]'
77                       '/span[2]/text()').extract_first()
78          # 房屋用途
79          i['house_usage'] = response.xpath
                        ('//div[@class="transaction"]//ul/li[4]'
80                       '/span[2]/text()').extract_first()
81          # 产权所有
82          i['property_own'] = response.xpath
                            ('//div[@class="transaction"]//ul/li[6]'
83                           '/span[2]/text()').extract_first()
84          # 图片url
```

```
85          i['images_urls'] = response.css('.smallpic >
                                        li::attr(data-pic)').extract()
86          yield i
```

（7）运行爬虫，运行完成之后查看数据。上面在 PyCharm 添加过 Mongo Plugin，现在通过插件连接 MongoDB，其中 label 栏位可以自行设置名称，如图 5.30 所示。

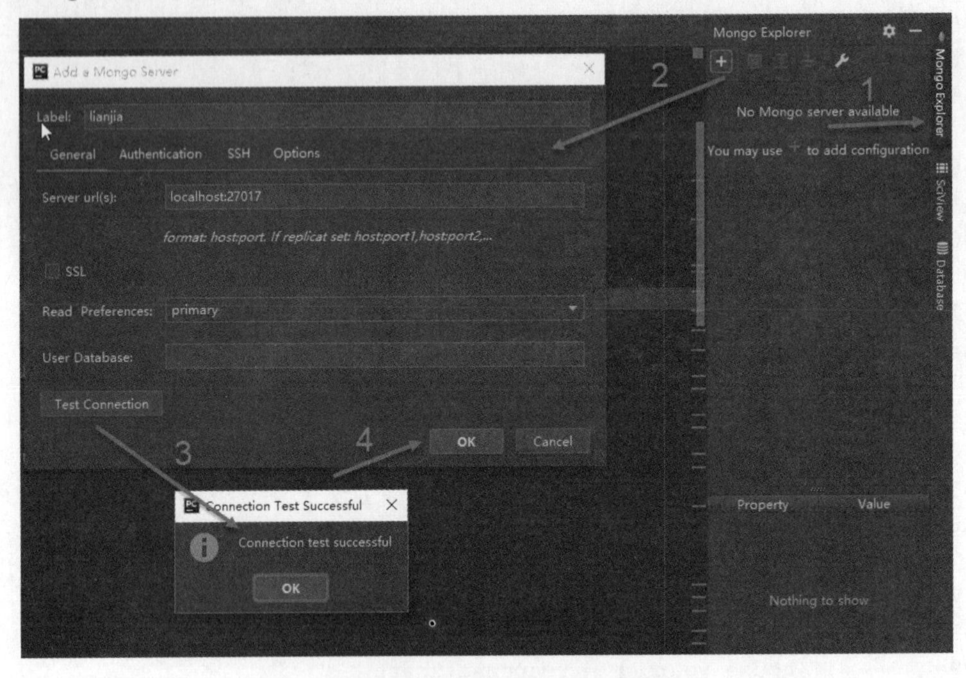

图 5.30　MongoPlugin 测试 MongoDB 连接

添加 MongoDB 服务器之后，右击它，选择 Connect to this server 菜单，连接服务器，之后会显示所有的数据库，如图 5.31 所示。

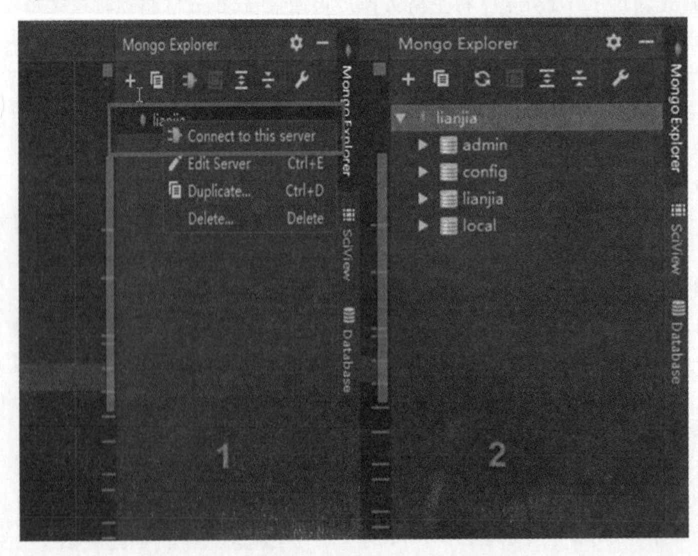

图 5.31　MongoPlugin 连接 MongoDB

双击数据库中的文档名称，查看抓取到的数据，文档右上角有两个按钮，其中左边按钮以键值对形式查看数据，右边按钮以表格形式查看数据，如图 5.32 所示。

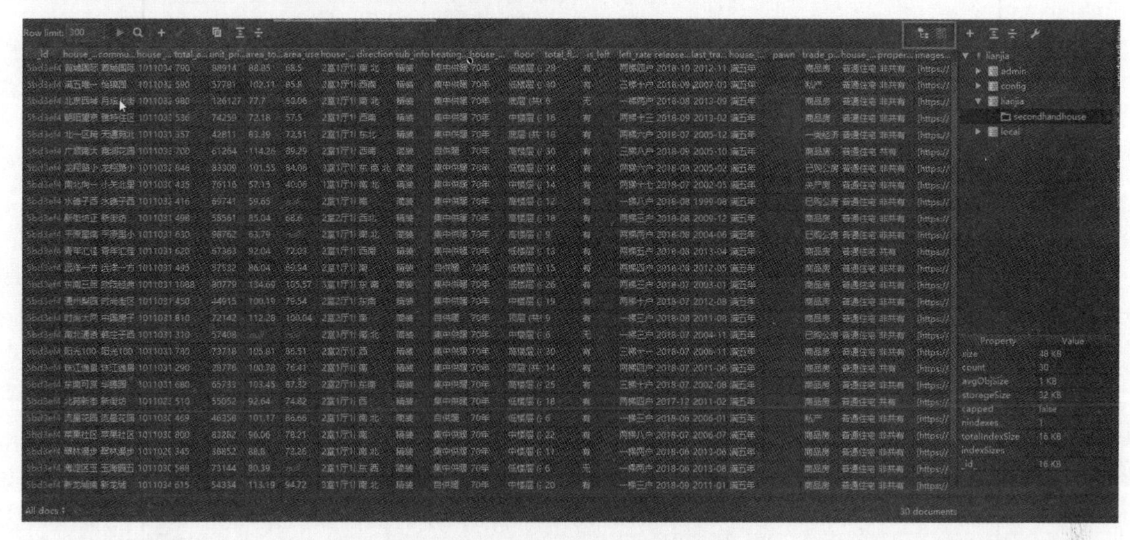

图 5.32　MongoPlugin 查看数据

再来查看保存的图片，在文件夹中可以看到，已经以二手房屋的标题作为文件夹名称存放了房屋的相关图片，如图 5.33 所示。

图 5.33　图片保存目录

文件夹中存储的图片如图 5.34 所示。

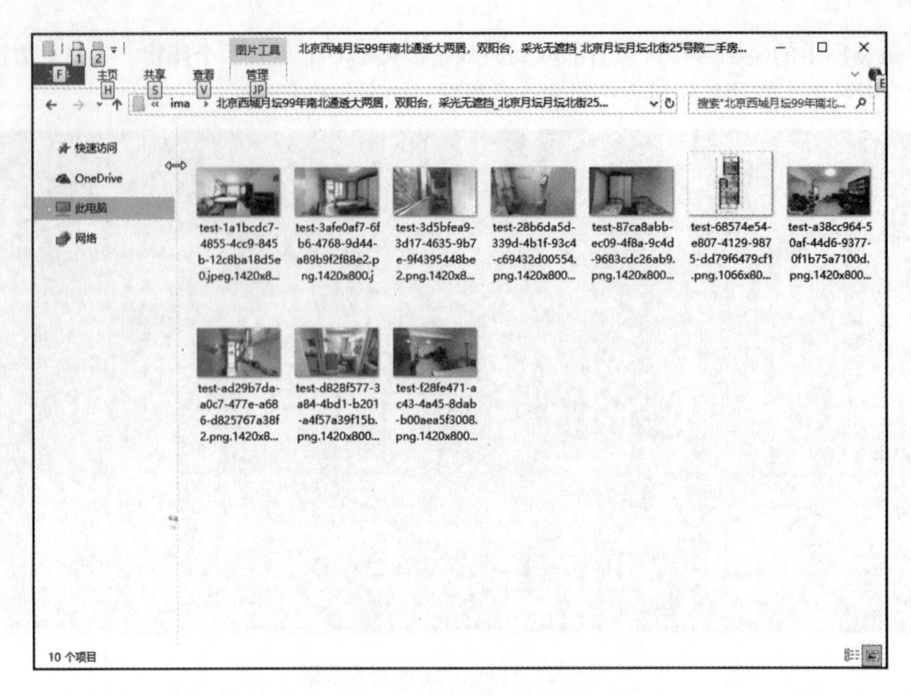

图 5.34　二手房对应图片

至此，一个使用数据库存储爬取数据的项目就结束了。本项目只抓取了北京地区二手房第一页的数据，读者如有兴趣可尝试将后续数据及其他地区的数据一并抓取下来。另外，请读者思考如何将数据存储到 MySQL 数据库。

第 6 章 Request 与 Response

Scrapy 框架中，有两个主要的对象 Request 与 Response，它们贯穿在爬虫的始终，最终爬虫通过它们将数据串联起来。

总的来说，Request 对象在 Spider 中生成，包含 HTTP 请求信息，在框架中经过一系列传递、处理，最终到达 Downloader 下载器，下载器执行 Request 中的请求进行数据抓取，将生成的响应包装成一个 Response 对象，再经过传递、处理，最终返回发送 Request 的 Spider，整个流程如图 6.1 所示。

图 6.1　传递 Response 与 Request

Request 类与 Response 类中都含有一些额外的功能，可以进行高度定制，本章会详细介绍。本章主要的知识点有：

- Request 类分析
- Request 中如何传递数据
- Response 类分析
- Response 子类介绍

6.1 Request 对象

Request 对象由 Spider 生成，包含指定的 HTTP 请求信息，最终传递到 Downloader 下载器进行数据下载。

6.1.1 Request 类详解

Request 类原型：

```
class scrapy.http.Request(url[, callback, method='GET', headers, body, 
cookies, meta, encoding='utf-8', priority=0, dont_filter=False, errback, flags])
```

参数说明：

- url(string)：请求 URL。
- callback(callable)：调用指定的方法处理请求生成的响应，如果不指定，就默认使用 parse()方法处理响应。需要注意的是，一旦发生错误，errback 指定的异常处理方法将会代替该方法被调用。
- method(string)：HTTP 请求方法，默认为 GET。
- meta(dict)：Request.meta 属性的初始值。
- body(str or unicode)：请求 body。如果值为 Unicode，那么将会用 encoding 指定的编码方式转化为 str 类型。如果 body 不指定，就会存储为空字符串，因此，无论 body 指定为何种类型，最终都是存储为一个字符串。
- headers(dict)：请求头。如果指定为 None，将不会发送请求头。
- cookies(dict or list)：请求 cookies，可以用以下两种方式发送：

  ① 使用字典发送：

```
request_with_cookies = Request(url="http://www.example.com",
                               cookies={'currency': 'USD', 'country': 'UY'})
```

  ② 使用字典列表发送：

```
request_with_cookies = Request(url="http://www.example.com",
                               cookies={'currency': 'USD', 'country': 'UY'})
request_with_cookies = Request(url=http://www.example.com,
cookies=[{'name': 'currency',
          'value': 'USD',
          'domain': 'example.com',
          'path': '/currency'}])
```

第 2 种形式支持自定义 cookies 中的 domain 与 path，这种情形通常是为了把 cookie 保存到之后的请求中，在接下来的请求中发送出去。

当一些网站返回 cookies 时，这些 cookies 将会存储在相对应的 cookies 域中，这是一个典型的浏览器行为。由于一些原因，当我们不想将接收到的 cookies 与现有的 cookies 合并时，我们可以指定 Request.meta 中的 dont_merge_cookies 字段值为 True，代码如下：

```
request_with_cookies = Request(url="http://www.example.com",
                               cookies={'currency': 'USD', 'country': 'UY'},
                                 meta={'dont_merge_cookies': True})
```

- encoding(string)：请求的编码，默认为 UTF-8。使用该指定的编码值对 URL 和 body 进行编码。
- priority(int)：请求的优先级，默认为 0。调度器依据优先级来处理请求的顺序。
- dont_filter(boolean)：表明调度器不需过滤该请求，默认为 False。当需要多次执行相同的请求时，可以指定为 True。在使用此参数时必须小心处理，否则会陷入死循环。
- errback(callable)：当处理请求抛出异常时调用该参数指定的方法。
- flags(list)：发送到请求的标志，可用于日志记录或类似目的。

Request 对象包含一些属性和方法，如表 6-1 所示。

表6-1　Request属性方法

属性或方法	说明
url	请求的 URL，注意这个属性值包含转义过的 URL，可以与原始 URL 不同，可用 replace()方法替换
method	HTTP 请求方式，如 GET、POST、PUT 等
headers	一个字典形式的请求头信息
body	包含请求 body 的字符串，可用 replace()方法替换
meta	包含在请求中的字典形式元数据。可以向其中添加在请求中会经常用到的属性，后面会进行详细讲解
copy()	根据原 Request 复制一个新的 Request
replace([url, method, headers, body, cookies, meta, encoding, dont_filter, callback, errback])	通过关键字替换相对应的参数，返回一个与原 Request 具有相同成员的 Request 对象

特别说明一下表 6-1 中的 Request.meta 键值。

新生成的 Request 的 meta 属性值为空，通常情况下根据启用的 Scrapy 组件进行填充。Request.meta 属性除了可以填充自定义数据外，还包含一些 Scrapy 和内置扩展可识别的特定键值，这些特定键值对 Request 起到特殊的限制作用，如表 6-2 所示。

表6-2　Request.meta属性

名称	说明
dont_redirect	True 或 False，若设定为 True，则禁止重定向
dont_retry	True 或 False，若设定为 True，则禁止重新抓取失败的网页
handle_httpstatus_list	列表，指定处理更多的 HTTP 返回状态码，默认处理 2xx 的状态码
handle_httpstatus_all	True 或 False，若指定为 True，则处理所有的 HTTP 返回码
dont_merge_cookies	True 或 False，若设定为 True，则新的 cookies 不与现有的 cookies 合并
cookiejar	用来管理 cookies，第 9 章模拟登录会详细介绍
dont_cache	True 或 False，若设定为 True，则禁止 HTTP 缓存
redirect_urls	list，允许重定向的 URL
bindaddress	执行外部请求时的 IP
dont_obey_robotstxt	True 或 False，设定为 True 时忽略网站 Robots 协议

（续表）

名称	说明
download_timeout	设置下载超时，单位为秒
download_maxsize	设置可下载的最大数据
download_latency	只读属性值，记录了发送请求到获得响应的时间
proxy	设定代理 URL
max_retry_times	设定最大失败重试次数

6.1.2 Request 回调函数与错误处理

当下载器处理完毕 Request 并生成 Response 时，调用回调函数，也就是 callback 指定的处理方法，调用的回调函数以该 Response 作为第 1 个参数。

【示例 6-1】Request 回调函数

```
01 def parse_page1(self, response):
02     return scrapy.Request("http://www.example.com/some_page.html",
03                           callback=self.parse_page2)
04
05 def parse_page2(self, response):
06     # this would log http://www.example.com/some_page.html
07     self.logger.info("Visited %s", response.url)
08
```

许多情况下，我们需要在不同的回调函数间传递参数，这时就可以使用 Request.meta 来处理参数的传递。例如我们收集用户信息的 Item 中，在第一个网页中只有用户名，而用户年龄来自另一个网页，可以如下处理：

【示例 6-2】传递参数

```
01 def parse_page1(self, response):
02     item = MyItem()
03     item['name'] = response.css('.name::text').extract_first()
04     request = scrapy.Request("http://www.example.com/some_page.html",
05                              callback=self.parse_page2)
06     request.meta['item'] = item
07     yield request
08
09 def parse_page2(self, response):
10     item = response.meta['item']
11     item['age'] = response.css('.age::text').extract_first()
12     yield item
```

当下载器处理 Request 抛出异常时，可以调用 errback 指定的异常处理方法。

errback 指定的异常处理方法接收一个 Twisted Failure 实例作为第 1 个参数，可用于追踪连接超时、DNS 错误等。先看如下示例：

【示例 6-3】异常捕获处理

```
01 import scrapy
02
03 from scrapy.spidermiddlewares.httperror import HttpError
04 from twisted.internet.error import DNSLookupError
05 from twisted.internet.error import TimeoutError, TCPTimedOutError
06
07 class ErrbackSpider(scrapy.Spider):
08     name = "errback_example"
09     start_urls = [
10         "http://www.httpbin.org/",              # 正常 HTTP 200 返回
11         "http://www.httpbin.org/status/404",    # 404 Not found error
12         "http://www.httpbin.org/status/500",    # 500 服务器错误
13         "http://www.httpbin.org:12345/",        # 超时无响应错误
14         "http://www.httphttpbinbin.org/",       # DNS 错误
15     ]
16
17     def start_requests(self):
18         for u in self.start_urls:
19             yield scrapy.Request(u, callback=self.parse_httpbin,
20                                     errback=self.errback_httpbin,
21                                     dont_filter=True)
22
23     def parse_httpbin(self, response):
24         self.logger.info('Got successful response from {}'.
                                                  format(response.url))
25         # 其他处理
26
27     def errback_httpbin(self, failure):
28         # 日志记录所有的异常信息
29         self.logger.error(repr(failure))
30
31         # 假设我们需要对指定的异常类型做处理，
32         # 我们需要判断异常的类型
33
34         if failure.check(HttpError):
35             # HttpError 由 HttpErrorMiddleware 中间件抛出
36             # 可以接收到非 200 状态码的 Response
37             response = failure.value.response
38             self.logger.error('HttpError on %s', response.url)
39
40         elif failure.check(DNSLookupError):
```

```
41            # 此异常由请求 Request 抛出
42            request = failure.request
43            self.logger.error('DNSLookupError on %s', request.url)
44
45        elif failure.check(TimeoutError, TCPTimedOutError):
46            request = failure.request
47            self.logger.error('TimeoutError on %s', request.url)
48
```

首先使用 import 在 scrapy 与 twisted 中引入了几种异常类型，在 start_url 中设定了一些异常的 URL。在 start_request 方法开始处理这些 URL 时，会抛出异常，调用 errback_httpbin()进行处理，failure 是一个 Twisted Failure 实例，failure.check()检查该异常类型是否在指定的参数列表中。运行爬虫，查看输出结果，如图 6.2 所示。

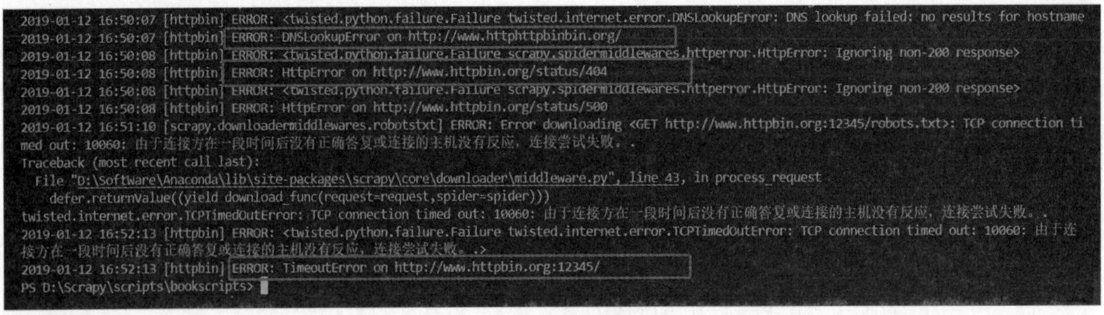

```
2019-01-12 16:50:07 [httpbin] ERROR: <twisted.python.failure.Failure twisted.internet.error.DNSLookupError: DNS lookup failed: no results for hostname
2019-01-12 16:50:07 [httpbin] ERROR: DNSLookupError on http://www.httphttpbinbin.org/
2019-01-12 16:50:08 [httpbin] ERROR: <twisted.python.failure.Failure scrapy.spidermiddlewares.httperror.HttpError: Ignoring non-200 response>
2019-01-12 16:50:08 [httpbin] ERROR: HttpError on http://www.httpbin.org/status/404
2019-01-12 16:50:08 [httpbin] ERROR: <twisted.python.failure.Failure scrapy.spidermiddlewares.httperror.HttpError: Ignoring non-200 response>
2019-01-12 16:50:08 [httpbin] ERROR: HttpError on http://www.httpbin.org/status/500
2019-01-12 16:51:10 [scrapy.downloadermiddlewares.robotstxt] ERROR: Error downloading <GET http://www.httpbin.org:12345/robots.txt>: TCP connection ti
med out: 10060: 由于连接方在一段时间后没有正确答复或连接的主机没有反应，连接尝试失败。.
Traceback (most recent call last):
  File "D:\SoftWare\Anaconda\lib\site-packages\scrapy\core\downloader\middleware.py", line 43, in process_request
    defer.returnValue((yield download_func(request=request,spider=spider)))
twisted.internet.error.TCPTimedOutError: TCP connection timed out: 10060: 由于连接方在一段时间后没有正确答复或连接的主机没有反应，连接尝试失败。.
2019-01-12 16:52:13 [httpbin] ERROR: <twisted.python.failure.Failure twisted.internet.error.TCPTimedOutError: TCP connection timed out: 10060: 由于连
接方在一段时间后没有正确答复或连接的主机没有反应，连接尝试失败。.>
2019-01-12 16:52:13 [httpbin] ERROR: TimeoutError on http://www.httpbin.org:12345/
PS D:\Scrapy\scripts\bookscripts>
```

图 6.2 爬虫结果

6.2 Response

Response 表示一个 HTTP 响应，通常由 Downloader 下载器生成，最终传递给 Spider 进行处理。Spider 接收 Response 之后，从中提取出数据，很重要的一点是，可以根据不同类型的数据选择相应的 Response 子类进行处理。

6.2.1 Response 类详解

Response 类原型：

```
    class scrapy.http.Response(url[, status=200, headers=None, body=b'',
flags=None, request=None])
```

参数说明：

- url(string)：响应的 URL。
- status(integer)：HTTP 响应码，默认为 200。
- headers(dict)：响应的头信息。

- body(bytes)：响应 body，可以使用 response.txt 转化为 string 类型进行访问。
- flags(list)：用来初始化 Response.flags。
- request(Request object)：初始化 Response.request，表示生成该 Response 的 Request。

同样，Response 也包含一些属性和方法，如表 6-3 所示。

表6-3 Response属性

属性或方法	说明
url	该响应的 url
status	响应码，如 200、404
headers	响应头信息，可以使用 get()获取指定头信息，如 response.headers.get('User-Agent')，或使用 getlist('name')获取指定名称的全部头信息
body	响应 body
request	生成 response 的 Request 对象。该属性是在所有请求和响应通过 Downloader 中间件时由 Scrapy 引擎指定。这意味着： （1）HTTP 重定向原始的请求指向重定向之后的响应。 （2）Response.request.url 并不总是等于 Response.url。 （3）此属性仅在 spider 代码及 Spider 中间件中可用
meta	这个值与 Response.request 的 meta 属性值一致。Response.meta 具有传播性，无论是重定向还是重试，都可以得到原始的 Request.meta 属性值，见示例【6-2】
flags	包含 response 标志的列表，flags 就是用于调试的标志
copy()	从当前 Response 复制一个新的 Response
replace([url, status, headers, body, request, flags, cls])	返回一个修改了指定参数的 Response 对象
urljoin(url)	使用 Response.url 与相对 url（相对地址）构成一个绝对 url（全地址）

6.2.2 Response 子类

下面介绍一些常用的 Response 子类，当然也可以自定义子类，添加指定功能。

1. TextResponse

TextResponse 原型：

```
class scrapy.http.TextResponse(url[, encoding[, ...]])
```

TextResponse 子类在 Response 基类的基础上添加了编码能力，因此适合用来处理二进制数据，如图像、音频、视频等媒体文件。

TextResponse 相比于 Response 添加了一个构造参数 encoding，其余参数与 Response 一致，这里仅对 encoding 参数进行说明。

- encoding(string)：一个包含用于处理此响应的编码类型的字符串。如果你创建一个带有 unicode 数据的 TextResponse 对象，此对象将使用这个指定的编码类型进行编码处理（注意响应的 body

属性是一个字符串）。如果 encoding 是 None（默认值），就在响应头信息和响应正文中查找编码类型进行使用。

TextResponse 对象还支持表 6-4 所示的属性和方法。

表6-4 TextResponse属性和方法

属性或方法	说明
text	响应体，unicode 类型数据。与 response.body.decode(response.encoding) 结果一致，会在内存中保存第一次的处理结果，无须每次转化，因此可以随时使用 response.text
encoding	一个表示该响应编码类型的字符串。该值通过以下方法顺序获取： （1）构造参数 encoding 指定的编码类型 （2）在响应头信息中 Content-Type HTTP 参数中指定的编码类型 （3）在 response 响应体中声明的编码类型 （4）通过 response 响应体推断编码类型
selector	与 response 中 selector 的使用方法一致，也支持 xpath 与 css 选择使用
follow(url, callback=None, method='GET', headers=None, body=None, cookies=None, meta=None, encoding=None, priority=0, dont_filter=False, errback=None)	根据 url 返回一个 Request 对象，用法与 Request 一致
body_as_unicode()	与 response.text 效果相同的一个方法

2. HtmlResponse

HtmlResponse 原型：

```
class scrapy.http.HtmlResponse(url[, ...])
```

HtmlResponse 是 TextResponse 的子类，只不过是添加了一个通过查看 HTML meta http-equiv 属性来自动添加编码的功能。

3. XMLResponse

XMLResponse 原型：

```
class scrapy.http.XmlResponse(url[, ...])
```

XmlResponse 也是 TextResponse 的一个子类，该类的自动添加编码功能来源于对 XML 声明行的检查。

第7章

Scrapy 中间件

在第 2 章介绍 Scrapy 时，从框架流程图中，我们了解到中间件包含两种类型：Spider（爬虫）中间件和 Downloader（下载器）中间件。这两个中间件分别对应处理 Response 与 Request 请求。

本章主要的知识点有：

- Spider 中间件介绍
- 如何编写 Spider 中间件
- Download 中间件介绍
- 如何编写 Download 中间件

7.1 编写自定义 Spider 中间件

从 2.10.2 小节图 2.22 展示的框架流程图中可以看到，Spider 中间件（Middleware）是介于 Scrapy 引擎与 Spider 中间的处理机制的钩子框架，可以添加代码来处理发送给 Spiders 的 Response 及 Spider 产生的 Item 和 Request。

7.1.1 激活中间件

启用 Spider 中间件时，需要将其加入 SPIDER_MIDDLEWARES 设置中。该设置位于 settings.py 文件中，是字典类型的，其中键为中间件的路径，值为中间件的顺序，代码如下：

```
SPIDER_MIDDLEWARES = {
    'myproject.middlewares.MySpiderMiddleware': 510,
}
```

自定义在 settings.py 中的中间件设置会与 Scrapy 内置的 SPIDER_MIDDWARES_BASE 设置合并（但不是覆盖），然后根据顺序（order）进行排序，最后得到启用中间件的有序列表：第一个中间件是最靠近引擎的，最后一个中间件是最靠近 Spider 的。

注意，关于如何分配中间件的顺序，请查看下面的 SPIDER_MIDDLEWARES_BASE 设置，而后根据要放置中间件的位置选择一个值。由于每个中间件执行不同的动作，自定义的中间件可能会依赖于之前（或者之后）执行的中间件，因此顺序是很重要的，如果不能确定自己的自定义中间件应该靠近哪个方向，就在 500～700 之间选择最为妥当。

Scrapy 内置 SPIDER_MIDDWARES_BASE：

```
'scrapy.spidermiddlewares.httperror.HttpErrorMiddleware': 50,
'scrapy.spidermiddlewares.offsite.OffsiteMiddleware': 500,
'scrapy.spidermiddlewares.referer.RefererMiddleware': 700,
'scrapy.spidermiddlewares.urllength.UrlLengthMiddleware': 800,
'scrapy.spidermiddlewares.depth.DepthMiddleware': 900,
```

如果想禁止内置的中间件，就必须在项目的 SPIDER_MIDDLEWARES 设置中定义该中间件，并将其值赋为 None。例如，想要关闭 off-site 中间件，可以设置：

```
SPIDER_MIDDLEWARES = {
    'myproject.middlewares.MySpiderMiddleware': 510,
    'scrapy.contrib.spidermiddleware.offsite.OffsiteMiddleware': None,
}
```

7.1.2 编写 Spider 中间件

编写 Spider 中间件与管道类似，每个中间件组件都实现了以下一个或多个方法的 Python 类。

1. process_spider_input(response,spider)

说明：

当参数 response 通过参数 spider 中间件时，该方法被调用，处理该 response，即在下载器中间件处理完成后，马上要进入某个回调函数 parse_xx()前被调用。

参数：

- response（Response 对象）：被处理的 Response。
- spider（Spider 对象）：该 Response 对应的 Spider。

返回值：

该方法应该返回 None 或者抛出一个异常。如果其返回 None，Scrapy 将会继续处理该 Response，调用所有其他的中间件直到 Spider 处理该 Response。如果其抛出一个异常（exception），Scrapy 将不会调用任何其他中间件的 process_spider_input()方法，而是调用 Request 的 errback。errback 的输出将会从另一个方向被重新输入中间件链中，使用 process_spider_output()方法来处理，当其抛出异常时调用 process_spider_exception()。

2. process_spider_output(response,result,spider)

说明：

当 Spider 处理完毕 Response 返回 result 时，即在爬虫运行 yield item 或者 yield scrapy.Request()的时候调用该方法。

参数：

- response（Response 对象）：生成该输出的 Response。
- result（包含 Request 或 Item 对象的可迭代对象）：Spider 返回的 result。
- spider（Spider 对象）：结果被处理的 Spider。

返回值：

该方法必须返回包含 Request 或 Item 对象的可迭代对象。如果是 Item，就会被交给 Pipeline；如果是 Request，就会被交给调度器，然后下载器中间件才会开始运行。

3. process_spider_exception(response,exception,spider)

说明：

当 spider 或（其他 spider 中间件的）process_spider_input()抛出异常时，该方法被调用。

参数：

- response（Response 对象）：异常抛出时处理的 Response。
- exception（Exception 对象）：抛出的异常。
- spider（Spider 对象）：抛出异常的 Spider。

返回值：

该方法必须返回 None 或者一个包含 Response 或 Item 对象的可迭代对象（iterable）。

如果返回 None，Scrapy 将继续处理该异常，调用中间件链中的其他中间件的 process_spider_exception()方法，直到所有中间件都被调用，该异常到达引擎（异常将被记录并被忽略）。

如果其返回一个可迭代对象，那么中间件链的 process_spider_output()方法将被调用，其他的中间件的 process_spider_exception()将不会被调用。

4. process_start_requests(start_requests, spider)

说明：

该方法以 Spider 启动的 Request 为参数被调用。

参数：

- start_requests（包含 Request 的可迭代对象）：start requests 列表。
- spider（Spider 对象）：启动 start requests 的 Spider。

返回值：

该方法接收的是一个可迭代对象（start_requests 参数），而且必须返回一个包含 Request 对象的可迭代对象。

7.2 Spider 内置中间件

Spider 已经内置了一些可以辅助爬取的中间件，通过这些中间件的启用、配置可以方便地进行爬虫的优化，提高爬取成功率与效率。

7.2.1 DepthMiddleware 爬取深度中间件

路径：

```
class scrapy.contrib.spidermiddleware.depth.DepthMiddleware
```

作用：

DepthMiddleware 是一个用于追踪被爬取网站中每个 Request 的爬取深度的中间件。深度是 start_urls 中定义 URL 的相对值，也就是相对 URL 的深度。例如定义 URL 为 http://www.example.com/article/，设置 DEPTH_LIMIT=1，那么限制爬取的只能是此 URL 下一级的网页。

它可用于限制爬取的最大深度，并根据深度控制请求优先级等。

DepthMiddleware 可以通过下列设置进行配置：

- DEPTH_LIMIT：爬取所允许的最大深度，如果为 0，就没有限制。
- DEPTH_STATS：是否收集爬取深度统计数据。
- DEPTH_PRIORITY：是否根据 Request 深度对其安排优先级。

7.2.2 HttpErrorMiddleware 失败请求处理中间件

路径：

```
class scrapy.contrib.spidermiddleware.httperror.HttpErrorMiddleware
```

作用：

过滤出所有失败（错误）的 HTTP Response，爬虫不需要消耗更多的资源，设置更为复杂的逻辑来处理这些异常 Request。根据 HTTP 标准，返回值在 200~300 的为成功的 Response。

如果想处理在这个范围之外的 Response，可以通过 Spider 的 handle_httpstatus_list 属性或 HTTPERROR_ALLOWED_CODES 设置来指定 Spider 能处理的 Response 返回值。

例如，如果想要处理返回值为 404 的 Response，那么可以这么做：

```
class MySpider(CrawlSpider):
    handle_httpstatus_list = [404]
```

Request.meta 中的 handle_httpstatus_list 键也可以用来指定每个请求所允许的响应码。也可以设置 handle_httpstatus_all 键值为 True 来处理任何响应状态码的请求。

除非必要且目的明确，否则不推荐处理非 200 状态码的响应。

HttpErrorMiddleware 部分设置项：

- HTTPERROR_ALLOWED_CODES：默认为[]，传递该列表中所有非 200 状态码的 Response。
- HTTPERROR_ALLOW_ALL：默认为 False，传递所有 Response，无论其状态码为何值。

7.2.3 OffsiteMiddleware 过滤请求中间件

路径：

```
class scrapy.contrib.spidermiddleware.offsite.OffsiteMiddleware
```

作用：

过滤掉所有 Spider 所覆盖主机域名外的 URL 请求。

该中间件过滤掉所有主机名不在 Spider 的 allowed_domains 属性值中的 Request。此属性中域名的子域名是被允许的，如果 allow_domains 中包含 www.example.com，那么允许 child.www.example.com，但不论是 www1.example.com 还是 example.com 都不允许。

当 Spider 返回一个主机名不在该 Spider 覆盖范围内的 Request 时，该中间件将会做一个类似于下面的日志记录：

```
DEBUG: Filtered offsite request to 'www.othersite.com': <GET
http://www.othersite.com/some/page.html>
```

为了避免在日志中记录太多无用信息，只会对每个新过滤的域名记录一次。例如，如果过滤出另一个 www.othersite.com 请求，就不会有新的记录。但如果过滤出 someothersite.com 请求，就会增加一条记录信息（第一次过滤时增加记录）。

如果 Spider 中没有定义 allowed_domains 属性，或该属性为空，此 offsite 中间件将不会过滤任何 Request。

如果 Request 设置了 dont_filter 属性，即使该 Request 的域名不在 allow_domain 属性值中，offsite 中间件也会允许该 Request。

7.2.4 RefererMiddleware 参考位置中间件

路径：

```
class scrapy.contrib.spidermiddleware.referer.RefererMiddleware
```

作用：

根据生成的 Response 的 URL 来填充 Request Referer 信息。

RefererMiddleware 有一个关键的设置项：REFERER_ENABLED，默认为 True，表示是否启用 referer 中间件。

7.2.5 UrlLengthMiddleware 网址长度限制中间件

路径：

```
class scrapy.contrib.spidermiddleware.urllength.UrlLengthMiddleware
```

作用：

过滤 URL 长度比 URLLENGTH_LIMIT 的值大的 Request。

UrlLengthMiddleware 有一个关键配置 URLLENGTH_LIMIT，表示允许爬取 URL 最长的长度。

7.3 编写自定义下载器中间件

下载器中间件是介于 Scrapy 的 Request 与 Response 中间的、用于处理请求与响应的钩子框架，它是一个轻量级的底层系统，用于全局修改 Scrapy 的请求与响应。

7.3.1 激活中间件

与激活 Spider 中间件一样，激活下载器中间件需要将其加入 DOWNLOADER_MIDDLEWARES 设置中，该设置位于 settings.py 文件中，是一个字典类型。其中，键为中间件的路径，值为中间件的执行顺序，代码如下：

```
DOWNLOADER_MIDDLEWARES = {
    'myproject.middlewares.CustomDownloaderMiddleware': 543,
}
```

自定义在 setting.py 中的下载器中间件会与 Scrapy 中定义 DOWNLOADER_MIDDLEWARES_BASE 设置相合并（并不会覆盖），然后根据顺序进行排序，最终得到一个启用的中间件有序列表。第一个中间件靠近 Scrapy 引擎，最后一个中间件靠近下载器。

下载器中间件的顺序同样很重要，由于每个中间件执行不同的动作，因此自定义的中间件可能会依赖于之前（或者之后）执行的中间件。参考以下 DOWNLOADER_MIDDLEWARES_BASE 顺序来设置自定义中间件的顺序：

```
    'scrapy.downloadermiddlewares.robotstxt.RobotsTxtMiddleware': 100,
    'scrapy.downloadermiddlewares.httpauth.HttpAuthMiddleware': 300,
    'scrapy.downloadermiddlewares.downloadtimeout.
DownloadTimeoutMiddleware': 350,
    'scrapy.downloadermiddlewares.defaultheaders.DefaultHeadersMiddleware':
400,
    'scrapy.downloadermiddlewares.useragent.UserAgentMiddleware': 500,
```

```
        'scrapy.downloadermiddlewares.retry.RetryMiddleware': 550,
        'scrapy.downloadermiddlewares.ajaxcrawl.AjaxCrawlMiddleware': 560,
        'scrapy.downloadermiddlewares.redirect.MetaRefreshMiddleware': 580,
        'scrapy.downloadermiddlewares.httpcompression.
HttpCompressionMiddleware': 590,
        'scrapy.downloadermiddlewares.redirect.RedirectMiddleware': 600,
        'scrapy.downloadermiddlewares.cookies.CookiesMiddleware': 700,
        'scrapy.downloadermiddlewares.httpproxy.HttpProxyMiddleware': 750,
        'scrapy.downloadermiddlewares.stats.DownloaderStats': 850,
        'scrapy.downloadermiddlewares.httpcache.HttpCacheMiddleware': 900,
```

如果想关闭在DOWNLOADER_MIDDLEWARES_BASE中定义的默认启用的中间件，就必须在项目的DOWNLOADER_MIDDLEWARES中设置该中间件的顺序值为None。例如，要禁用user-agent中间件：

```
    DOWNLOADER_MIDDLEWARES = {
        'myproject.middlewares.CustomDownloaderMiddleware': 543,
        'scrapy.downloadermiddlewares.useragent.UserAgentMiddleware': None,
    }
```

7.3.2 编写下载器中间件

编写下载器中间件与Spider中间件一样，都是实现以下一个或几个方法的Python类：

1. process_request(request, spider)

说明：

当Request经过下载器中间件时调用该方法。

参数：

- request（Request对象）：处理的Request。
- spider（Spider对象）：该Request对应的Spider。

返回值：

该方法必须返回None、Response对象、Request对象或IgnoreRequest异常其中之一。

如果其返回None，Scrapy将执行其他中间件相应的方法继续处理该Request，直到合适的下载器处理函数被调用，该Request被执行处理（对应的Response被下载）。

如果返回Response对象，Scrapy将不再调用其他中间件的process_request()或process_exception()方法，或相应地下载函数，并将返回该Response。已启用的中间件的process_response()方法会在每个Response返回时被调用。

如果返回Request对象，Scrapy会停止调用process_request()方法，并重新调度处理返回的Request。当新返回的Request被执行后，将会根据下载的Response调用相应的中间件处理。

如果其抛出一个 IgnoreRequest 异常，已启用下载中间件的 process_exception()方法会被调用。如果没有任何一个中间件处理该异常，Request.errback 会被调用。如果没有代码处理抛出的异常，那么该异常被忽略且不记录到日志中。

2. process_response(request,response,spider)

说明：

当 Response 经过下载器中间件时调用该方法。

参数：

- request（Request 对象）：Response 所对应的 Request。
- response（Response 对象）：被处理的 Response。
- spider（Spider 对象）：Response 所对应的 Spider。

返回值：

该方法必须返回 Response 对象、Request 对象或 IgnoreRequest 异常其中之一。

如果其返回一个 Response（可以与处理的 Response 相同，也可以是全新的对象），那么该 Response 会被其他中间件的 process_response()方法处理。

如果其返回一个 Request 对象，那么中间件停止处理，返回的 Request 会被重新调度下载。该处理类似于 process_request()处理返回的 Request 的步骤。

如果其抛出一个 IgnoreRequest 异常，那就调用 Request.errback。如果没有代码处理抛出的异常，那么该异常被忽略且不记录在日志中。

3. process_exception(request, exception, spider)

说明：

当下载处理器（download handler）或 process_request()（下载中间件）抛出异常（包括 IgnoreRequest 异常）时，Scrapy 调用 process_exception()。

参数：

- request（Request 对象）：产生异常的 Request。
- exception（Exception 对象）：抛出的异常。
- spider（Spider 对象）：Request 对应的 Spider。

返回值：

该方法必须返回 None、Response 对象或 Request 对象其中之一。

如果其返回 None，Scrapy 将会继续处理该异常，接着调用已启用的其他中间件的 process_exception()方法，直到所有中间件都被调用完毕，并且默认异常处理执行。

如果其返回一个 Response 对象，就调用已启用中间件的 process_response()方法。Scrapy 将不会调用任何其他中间件的 process_exception()方法。

如果其返回一个 Request 对象，就返回的 Request 将会被重新调用下载。这将停止执行中间件的 process_exception()方法。

7.4　下载器内置中间件

同样，下载器内置了一些非常有用的中间件，可以方便地配置下载参数，如 Cookie、代理等。下面介绍一些常用的中间件。

7.4.1　CookiesMiddleware

路径：

```
class scrapy.contrib.downloadermiddleware.cookies.CookiesMiddleware
```

作用：

该中间件可以使用 Cookie 爬取网站数据。记录了向 Web Server 发送的 Cookie，并在之后的 Request 请求中发送回去，就像操作浏览器一样。

CookiesMiddleware 设置项：

- COOKIES_ENABLE：默认为 True，该配置指定是否启用 cookies 中间件，如果指定为 False，就不会使用 cookies。需要注意的是，如果 Request.meta 参数 dont_merge_cookies 指定为 True，那么无论 COOKIES_ENABLE 指定为何值，Request 都不会向服务器传递任何 cookies，Response 接收到的 cookies 也不会做任何处理。
- COOKIES_DEBUG：默认为 False，如果该参数指定为 True，那么将会记录所有的请求发送的 cookies 和响应接收到的 cookies。

很有用的一点是，每个蜘蛛爬虫可以保存多个 cookies，只需要为 Request.meta 指定 cookiejar 值。如下所示，通过传递不同的标识符使用不同的 Cookie：

```
for i, url in enumerate(urls):
    yield scrapy.Request(url, meta={'cookiejar': i},
        callback=self.parse_page)
```

需要注意的是，在 Request.meta 中 cookiejar 是没有黏性的。上例中，在调用 parse_page 时，需要将上次请求的 cookiejar 传递过来才能继续使用：

```
def parse_page(self, response):
    # do some processing
    return scrapy.Request("http://www.example.com/otherpage",
        meta={'cookiejar': response.meta['cookiejar']},
        callback=self.parse_other_page)
```

7.4.2 HttpProxyMiddleware

路径：

```
class scrapy.downloadermiddlewares.httpproxy.HttpProxyMiddleware
```

作用：

此中间件可以通过在 Request.meta 中添加 proxy 属性值为该请求设置 HTTP 代理，默认获取代理的方法是通过以下环境变量来获取代理地址：

- http_proxy
- https_proxy
- no_proxy

在 settings.py 中的设置：

- HTTPPROXY_ENABLED：默认为 False，表示是否激活 HttpProxyMiddleware。
- HTTPPROXY_AUTH_ENCODING：代理有验证时的账户信息编码方式。

7.5 实战：为爬虫添加中间件

【示例 7-1】为前面的爬虫项目添加中间件

了解爬虫中间件与下载器中间件之后，我们为第 5 章的项目分别添加一个 Spider 中间件与 Downloader 中间件。middlewares.py 代码如下：

```
01 # -*- coding: utf-8 -*-
02
03 # Define here the models for your spider middleware
04 #
05 # See documentation in:
06 # https://doc.scrapy.org/en/latest/topics/spider-middleware.html
07
08 from scrapy import signals
09 import scrapy
10 import random
11
12
13 class LianjiaSpiderMiddleware(object):
14     """
15     利用 Scrapy 数据收集功能记录相同小区的数量
16     """
17     def __init__(self, stats):
18         self.stats = stats
```

```
19
20     @classmethod
21     def from_crawler(cls, crawler):
22        return cls(stats=crawler.stats)
23
24     def process_spider_output(self, response, result, spider):
25        """
26        从item中获取小区名称，在数据收集中记录相同小区数量
27        :param response:
28        :param result:
29        :param spider:
30        :return:
31        """
32        for item in result:
33           if isinstance(item,scrapy.Item):
34              # 从result中的item获取小区名称
35              community_name = item['community_name']
36              # 在数据统计中为相同的小区增加数量值
37              self.stats.inc_value(community_name)
38           yield item
39
40
41  class LianjiaDownloaderMiddleware(object):
42     """
43     为请求添加代理
44     """
45     def __init__(self,proxy_list):
46        self.proxy_list = proxy_list
47
48     @classmethod
49     def from_crawler(cls, crawler):
50        # 从settings.py中获取代理列表
51        return cls(
52           proxy_list=crawler.settings.get('PROXY_LIST')
53        )
54
55     def process_request(self, request, spider):
56        # 从代理列表中随机选取一个添加至请求
57        proxy = random.choice(self.proxy_list)
58        request.meta['proxy'] = proxy
59
60     def spider_opened(self, spider):
61        spider.logger.info('Spider opened: %s' % spider.name)
```

在 LianjiaSpiderMiddleware 中，我们使用了 Scrapy 的数据统计功能，此功能在 8.3 节会详细介绍，读者只需知道在本章代码中可以统计指定对象的次数即可。

在 LianjiaDownloaderMiddleware 中，我们从 settings.py 中获取代理列表，然后随机选取一个添加至请求中，代理列表读者可以从网上获取，有条件的读者可以使用付费代理，连接起来会更稳定。setting.py 修改如下：

```
01 # Enable or disable spider middlewares
02 # See https://doc.scrapy.org/en/latest/topics/spider-middleware.html
03 SPIDER_MIDDLEWARES = {
04    'lianjia.middlewares.LianjiaSpiderMiddleware': 543,
05 }
06
07 # Enable or disable downloader middlewares
08 # See https://doc.scrapy.org/en/latest/topics/
                                        downloader-middleware.html
09 DOWNLOADER_MIDDLEWARES = {
10    'lianjia.middlewares.LianjiaDownloaderMiddleware': 543,
11 }
12
13
14 PROXY_LIST = [
15 'http://116.209.57.41:9999',
16 'http://117.90.252.151:9999',
17 'http://221.239.86.26:32228',
18 'http://117.95.12.239:9999',
19 'http://18.223.141.123:80',
20 'http://121.232.148.113:9000',
21 'http://120.198.230.65:8080',
22 'http://113.122.168.105:9999',
23 'http://218.95.48.156:9000',
24 'http://115.223.207.109:9000',
25 'http://183.3.221.186:8118',
26 'http://114.234.81.72:9000',
27 'http://111.177.177.87:9999',
28 'http://60.217.64.237:45091',
29 'http://36.248.129.240:9999'
30 ]
```

运行结果如图 7.1 所示，可以看到每一个小区出现的次数都被统计到。

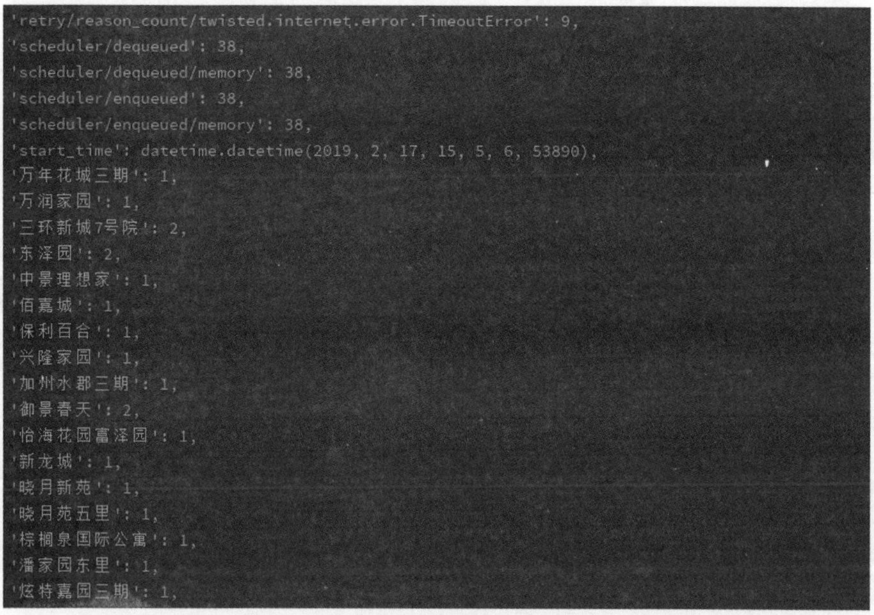

```
'retry/reason_count/twisted.internet.error.TimeoutError': 9,
'scheduler/dequeued': 38,
'scheduler/dequeued/memory': 38,
'scheduler/enqueued': 38,
'scheduler/enqueued/memory': 38,
'start_time': datetime.datetime(2019, 2, 17, 15, 5, 6, 53890),
'万年花城三期': 1,
'万润家园': 1,
'三环新城7号院': 2,
'东泽园': 2,
'中景理想家': 1,
'佰嘉城': 1,
'保利百合': 1,
'兴隆家园': 1,
'加州水郡三期': 1,
'御景春天': 2,
'恰海花园富泽园': 1,
'新龙城': 1,
'晓月新苑': 1,
'晓月苑五里': 1,
'棕榈泉国际公寓': 1,
'潘家园东里': 1,
'炫特嘉园三期': 1,
```

图 7.1　中间件统计信息

第8章 Scrapy配置与内置服务

Scrapy 框架中可以通过一系列的配置来定制组件，包括核心（Core）、插件（Extension）、管道（Pipeline）及 Spider 组件。同时不同的内置服务也需要在配置中进行设定，如邮件服务等。进行合理的配置才能使 Scrapy 正常工作。

本章主要的知识点有：

- Scrapy 基本配置
- Scrapy 常用服务

8.1 Scrapy 配置简介

配置的基础结构提供了键值映射方式的全局命名空间，也就是说可以从任何位置访问这些属性。通过代码可以从中提取配置值。可以通过不同的机制来设定配置信息。本节将介绍 Scrapy 基本配置。

8.1.1 命令行选项（优先级最高）

执行命令行命令时，如果在项目内执行，默认就使用项目内配置，即 settings.py 中的配置信息，如果在项目外执行，就使用默认的命令行配置。不过在任何地方执行命令行命令，都可以使用命令行参数-s（或--set）来覆盖一个（或更多）配置信息，因为命令行的参数配置具有最高的优先级，命令如下：

```
scrapy crawl myspider -s LOG_FILE=scrapy.log
```

8.1.2　每个爬虫内配置

项目内每一个编写的爬虫默认使用的都是项目配置，即 settings.py 中的配置。但我们仍然可以为每一个爬虫设定不同的配置，比如有些爬虫需要调用某些中间件，而有些爬虫则不需要，只需要设定 custom_settings 即可，示例如下。

```
01 import scrapy
02
03 class MySpider(scrapy.Spider):
04     name = 'myspider'
05
06     custom_settings = {
07         'SOME_SETTING': 'some value',
08     }
```

8.1.3　项目设置模块

通过 scrapy startproject 命令创建项目之后，都会生成一个 settings.py 文件，该文件就是该项目的配置文件，默认的配置如下：

```
01 # -*- coding: utf-8 -*-
02
03 # Scrapy settings for segmentfault project
04 #
05 # For simplicity, this file contains only settings considered important or
06 # commonly used. You can find more settings consulting the documentation:
07 #
08 #     https://doc.scrapy.org/en/latest/topics/settings.html
09 #     https://doc.scrapy.org/en/latest/topics/downloader-middleware.html
10 #     https://doc.scrapy.org/en/latest/topics/spider-middleware.html
11
12 # Scrapy 项目名字
13 BOT_NAME = 'segmentfault'
14
15 # Scrapy 搜索 spider 的模块列表
16 SPIDER_MODULES = ['segmentfault.spiders']
17 # 使用爬虫创建命令 genspider 创建爬虫时生成的模块
18 NEWSPIDER_MODULE = 'segmentfault.spiders'
19
20
21 # 默认的 USER_AGENT，使用 BOT_NAME 配置生成，建议覆盖
22 #USER_AGENT = 'segmentfault (+http://www.yourdomain.com)'
23
```

```
24 # 如果启用，Scrapy 则会遵守网站 Rebots.txt 协议，建议设置为 False
25 ROBOTSTXT_OBEY = True
26
27 # 配置 Scrapy 最大并发数，默认为 32，一般需要增大设置
28 #CONCURRENT_REQUESTS = 32
29
30 # 为同一个站点设置下载延迟
31 # See https://doc.scrapy.org/en/latest/topics/settings.html#download-delay
32 # See also autothrottle settings and docs
33 #DOWNLOAD_DELAY = 3
34 # 下载延迟的设置只会根据以下两个中的一个生效
35 # 对单个网站设置最大的请求并发数
36 #CONCURRENT_REQUESTS_PER_DOMAIN = 16
37 # 对单个 IP 设置最大的请求并发数
38 #CONCURRENT_REQUESTS_PER_IP = 16
39
40 # 禁用 Cookie，默认 True 启用，建议为 False
41 #COOKIES_ENABLED = False
12 # 关闭 Telent 控制台，默认启用
43 #TELNETCONSOLE_ENABLED = False
44
45 # 默认的请求头，根据爬取网站覆盖
46 #DEFAULT_REQUEST_HEADERS = {
47 #   'Accept': 'text/html,application/xhtml+xml,application/xml;q=0.9,
                */*;q=0.8',
48 #   'Accept-Language': 'en',
49 #}
50
51 # 启用 Spider 爬虫中间件
52 # See https://doc.scrapy.org/en/latest/topics/spider-middleware.html
53 #SPIDER_MIDDLEWARES = {
54 #    'segmentfault.middlewares.SegmentfaultSpiderMiddleware': 543,
55 #}
56
57 # 启用 Downloader 下载器中间件
58 # See https://doc.scrapy.org/en/latest/topics/downloader-middleware.html
59 #DOWNLOADER_MIDDLEWARES = {
60 #    'segmentfault.middlewares.SegmentfaultDownloaderMiddleware': 543,
61 #}
62
63 # 启用扩展
64 # See https://doc.scrapy.org/en/latest/topics/extensions.html
65 #EXTENSIONS = {
66 #    'scrapy.extensions.telnet.TelnetConsole': None,
67 #}
```

```
68
69 # 配置管道信息
70 # See https://doc.scrapy.org/en/latest/topics/item-pipeline.html
71 #ITEM_PIPELINES = {
72 #    'segmentfault.pipelines.SegmentfaultPipeline': 300,
73 #}
74
75 # 启用配置 AutoThrottle 扩展，默认禁用，建议启用
76 # See https://doc.scrapy.org/en/latest/topics/autothrottle.html
77 #AUTOTHROTTLE_ENABLED = True
78 # 初始化下载延迟
79 #AUTOTHROTTLE_START_DELAY = 5
80 # 高延迟下最大的下载延迟
81 #AUTOTHROTTLE_MAX_DELAY = 60
82 # Scrapy 请求应该并行发送每个远程服务器的平均数量
83 #AUTOTHROTTLE_TARGET_CONCURRENCY = 1.0
84 # 启用调试模式，统计每一个响应状态数据
85 #AUTOTHROTTLE_DEBUG = False
86
87 # 启用和配置 HTTP 缓存
88 # See https://doc.scrapy.org/en/latest/topics/
           downloader-middleware.html#httpcache-middleware-settings
89 #HTTPCACHE_ENABLED = True
90 # HTTP 缓存过期时间
91 #HTTPCACHE_EXPIRATION_SECS = 0
92 # HTTP 缓存目录
93 #HTTPCACHE_DIR = 'httpcache'
94 # HTTP 缓存忽略的响应状态码
95 #HTTPCACHE_IGNORE_HTTP_CODES = []
96 # HTTP 缓存存储目录
97 #HTTPCACHE_STORAGE = 'scrapy.extensions.httpcache.FilesystemCacheStorage'
```

settings.py 配置文件很重要，当我们编写完管道、中间件，使用扩展时，一定要在 settings.py 中启用，才能够正常使用。

8.1.4 默认的命令行配置

在项目内使用命令行时，默认的配置为项目配置，在项目外使用命令行时，默认使用 Scrapy 全局配置。

在项目中使用：

```
>scrapy settings --get BOT_NAME
scrapybot
```

在项目外使用：

```
import scrapy
>scrapy settings --get BOT_NAME
scrapybot
```

8.1.5 默认全局配置（优先级最低）

Scrapy 默认全局配置是所有配置信息的基础，每种配置都是在此基础上进行覆盖的。可以通过 scrapy.settings.default_settings 访问全局配置。由于全局配置的项目很多，因此这里不一一介绍，读者可参考 Scrapy 官方文档进行查看（https://docs.scrapy.org/en/latest/topics/settings.html）。

8.2 日　　志

日志是查看程序运行状态的主要方法，特别是程序运行出错时，主要根据日志来检查错误，进行修正。Scrapy 使用 Python 内置的日志系统记录事件日志，在使用日志功能之前需要先进行一些配置。

- LOG_FILE：指定日志文件，如果为 None，就使用标准错误输出。
- LOG_ENABLED：是否启用日志，为 True 时启用日志，为 False 时不启用。
- LOG_ENCODING：使用指定的编码方式输出日志，默认为 UTF-8。
- LOG_LEVEL：日志记录的最低级别。可选的级别有 CRITICAL、ERROR、WARNING、INFO、DEBUG。默认为 DEBUG，打印所有记录。
- LOG_FORMAT：日志输出格式，默认为'%(asctime)s [%(name)s] %(levelname)s: %(message)s'。
- LOG_DATEFORMAT：日志日期记录格式，默认格式为'%Y-%m-%d %H:%M:%S'。
- LOG_STDOUT：默认为 False，如果为 True，那么表示进程所有的标准输出（及错误）将被重定向到 log 中。例如执行 print 'hello'，其将会在 Scrapy log 中显示。

Log 的使用非常简单：

```
import scrapy
>>> import logging
>>> logger=logging.getLogger('LogTest')
>>> logger.warning('warning message')
WARNING:LogTest:warning message
>>>
```

在每一个 Scrapy 爬虫实例中都提供了一个可以直接使用的 logger，代码如下：

```
01 import scrapy
02
03 class MySpider(scrapy.Spider):
```

```
04
05      name = 'myspider'
06      start_urls = ['http://quotes.toscrape.com']
07
08      def parse(self, response):
09         self.logger.info('Parse function called on %s', response.url)
```

输出结果如图 8.1 所示。

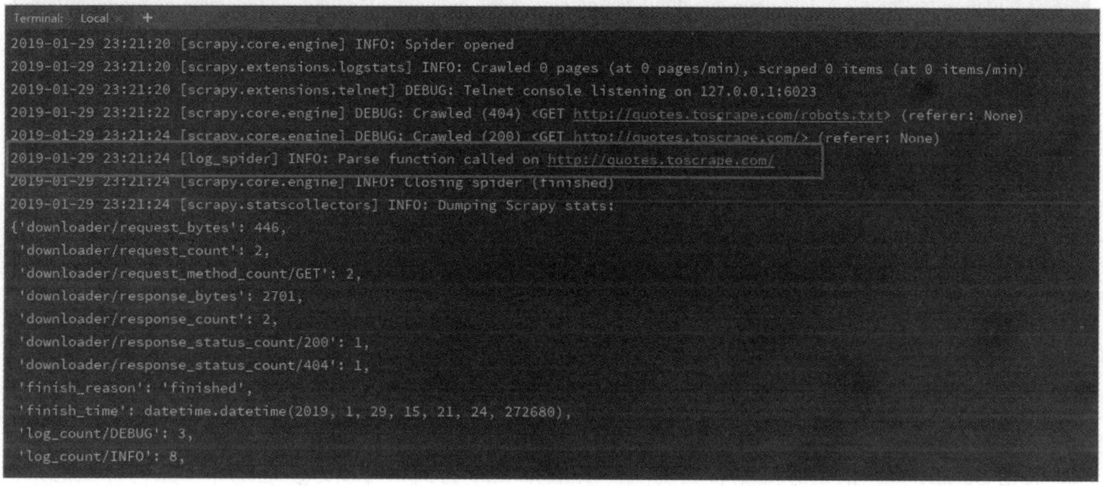

```
Terminal:  Local  +
2019-01-29 23:21:20 [scrapy.core.engine] INFO: Spider opened
2019-01-29 23:21:20 [scrapy.extensions.logstats] INFO: Crawled 0 pages (at 0 pages/min), scraped 0 items (at 0 items/min)
2019-01-29 23:21:20 [scrapy.extensions.telnet] DEBUG: Telnet console listening on 127.0.0.1:6023
2019-01-29 23:21:22 [scrapy.core.engine] DEBUG: Crawled (404) <GET http://quotes.toscrape.com/robots.txt> (referer: None)
2019-01-29 23:21:24 [scrapy.core.engine] DEBUG: Crawled (200) <GET http://quotes.toscrape.com/> (referer: None)
2019-01-29 23:21:24 [log_spider] INFO: Parse function called on http://quotes.toscrape.com/
2019-01-29 23:21:24 [scrapy.core.engine] INFO: Closing spider (finished)
2019-01-29 23:21:24 [scrapy.statscollectors] INFO: Dumping Scrapy stats:
{'downloader/request_bytes': 446,
 'downloader/request_count': 2,
 'downloader/request_method_count/GET': 2,
 'downloader/response_bytes': 2701,
 'downloader/response_count': 2,
 'downloader/response_status_count/200': 1,
 'downloader/response_status_count/404': 1,
 'finish_reason': 'finished',
 'finish_time': datetime.datetime(2019, 1, 29, 15, 21, 24, 272680),
 'log_count/DEBUG': 3,
 'log_count/INFO': 8,
```

图 8.1　记录日志

默认的 logger 使用的是爬虫名称，也可以自己指定名称：

```
01  import logging
02  import scrapy
03
04  logger = logging.getLogger('custom_logger')
05
06  class MySpider(scrapy.Spider):
07
08      name = 'myspider'
09      start_urls = [' http://quotes.toscrape.com ']
10
11      def parse(self, response):
12         logger.info('Parse function called on %s', response.url)
```

输出结果如图 8.2 所示。

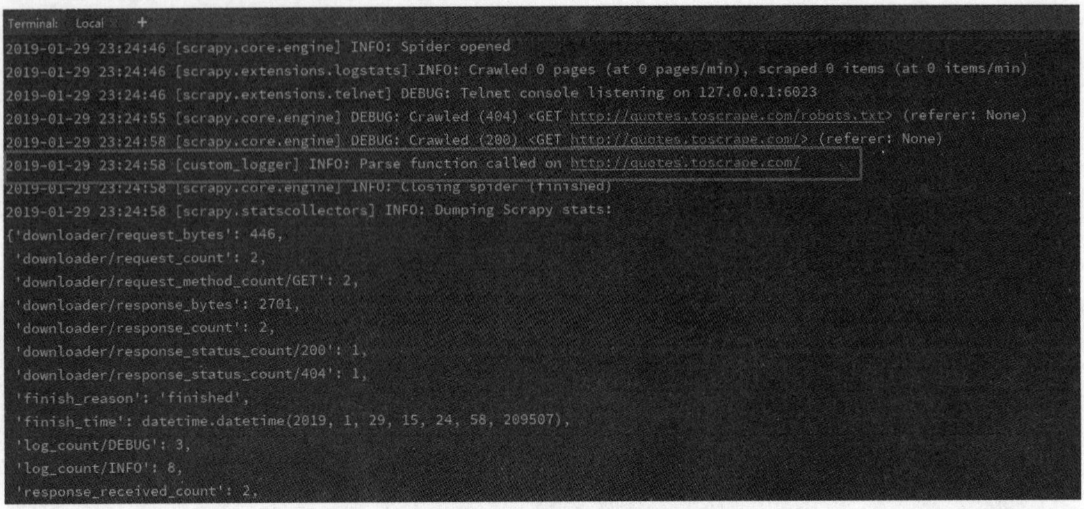

```
Terminal:  Local  +
2019-01-29 23:24:46 [scrapy.core.engine] INFO: Spider opened
2019-01-29 23:24:46 [scrapy.extensions.logstats] INFO: Crawled 0 pages (at 0 pages/min), scraped 0 items (at 0 items/min)
2019-01-29 23:24:46 [scrapy.extensions.telnet] DEBUG: Telnet console listening on 127.0.0.1:6023
2019-01-29 23:24:55 [scrapy.core.engine] DEBUG: Crawled (404) <GET http://quotes.toscrape.com/robots.txt> (referer: None)
2019-01-29 23:24:58 [scrapy.core.engine] DEBUG: Crawled (200) <GET http://quotes.toscrape.com/> (referer: None)
2019-01-29 23:24:58 [custom_logger] INFO: Parse function called on http://quotes.toscrape.com/
2019-01-29 23:24:58 [scrapy.core.engine] INFO: Closing spider (finished)
2019-01-29 23:24:58 [scrapy.statscollectors] INFO: Dumping Scrapy stats:
{'downloader/request_bytes': 446,
 'downloader/request_count': 2,
 'downloader/request_method_count/GET': 2,
 'downloader/response_bytes': 2701,
 'downloader/response_count': 2,
 'downloader/response_status_count/200': 1,
 'downloader/response_status_count/404': 1,
 'finish_reason': 'finished',
 'finish_time': datetime.datetime(2019, 1, 29, 15, 24, 58, 209507),
 'log_count/DEBUG': 3,
 'log_count/INFO': 8,
 'response_received_count': 2,
```

图 8.2　指定 logger 名称

8.3　数据收集

利用 Scrapy 提供的统计数据收集功能，以 key/value 方式，可以方便地统计一些特殊信息，包括指定数据的统计，比如特定的关键词、404 页面等。Scrapy 提供的这种收集数据机制叫作 Stats Collection。

数据收集器对每个 Spider 保持一个状态表。当 Spider 启动时，该表自动打开；当 Spider 关闭时，该表自动关闭。

通过 stats 属性来使用数据收集器。下面是在扩展中使用状态的例子：

```
01 class ExtensionThatAccessStats(object):
02 
03     def __init__(self, stats):
04         self.stats = stats
05 
06     @classmethod
07     def from_crawler(cls, crawler):
08         return cls(crawler.stats)
```

stats 属性有以下属性值可以配置。

- 设置数据：

```
stats.set_value('hostname', socket.gethostname())
```

- 增加数据值：

```
stats.inc_value('pages_crawled')
```

- 当新的值比原来的值大时设置数据：

```
stats.max_value('max_items_scraped', value)
```

- 当新的值比原来的值小时设置数据：

```
stats.min_value('min_free_memory_percent', value)
```

- 获取数据：

```
>>> stats.get_value('pages_crawled')
8
```

- 获取所有数据：

```
stats.get_stats()
{'pages_crawled': 1238, 'start_time': datetime.datetime(2009, 7, 14, 21, 47, 28, 977139)}
```

【示例 8-1】数据收集使用，统计名人名言网站中（http://quotes.toscrape.com/）标签为 love 的名言数量

（1）创建项目：

```
>>>scrapy startproject tagcount
```

（2）创建爬虫：

```
>>>cd tagcount
>>>scrapy genspider tags quotes.toscrape.com
```

（3）元素定位分析与示例 4-1 一致，只需添加数据收集代码，爬虫文件 tags.py 代码如下：

```
01 # -*- coding: utf-8 -*-
02 import scrapy
03 from scrapy import Request
04 from tagcount.items import TagcountItem
05
06
07 class TagsSpider(scrapy.Spider):
08     name = 'tags'
09     allowed_domains = ['quotes.toscrape.com']
10     start_urls = ['http://quotes.toscrape.com/']
11
12     def parse(self, response):
13         quotes = response.css('.quote')
14         for quote in quotes:
15             item = TagcountItem()
16             # 提取内容数据
17             item['author'] = quote.css('.author::text').extract_first()
18             item['content'] = quote.css('.text::text').extract_first()
```

```
19            item['tag'] = quote.css('.tag::text').extract()
20            if 'love' in item['tag']:
21                # 如果"love"在获取的 tag 内容中，则"love"统计数量+1
22                self.crawler.stats.inc_value('love')
23            yield item
24
25        next_page = response.css('.next > a::attr(href)').extract_first()
26        if next_page is not None:
27            yield Request(response.urljoin(next_page),
                            callback=self.parse)
28
```

其中，self.crawler.stats.inc_value('love')每当检测到“love”时增加“love”标签的数量统计值。

（4）编写 items.py 文件如下：

```
01 import scrapy
02
03 class QuotesItem(scrapy.Item):
04     # define the fields for your item here like:
05     # name = scrapy.Field()
06     author = scrapy.Field()
07     content = scrapy.Field()
08     tag = scrapy.Field()
```

（5）运行爬虫文件，输出结果如图 8.3 所示。

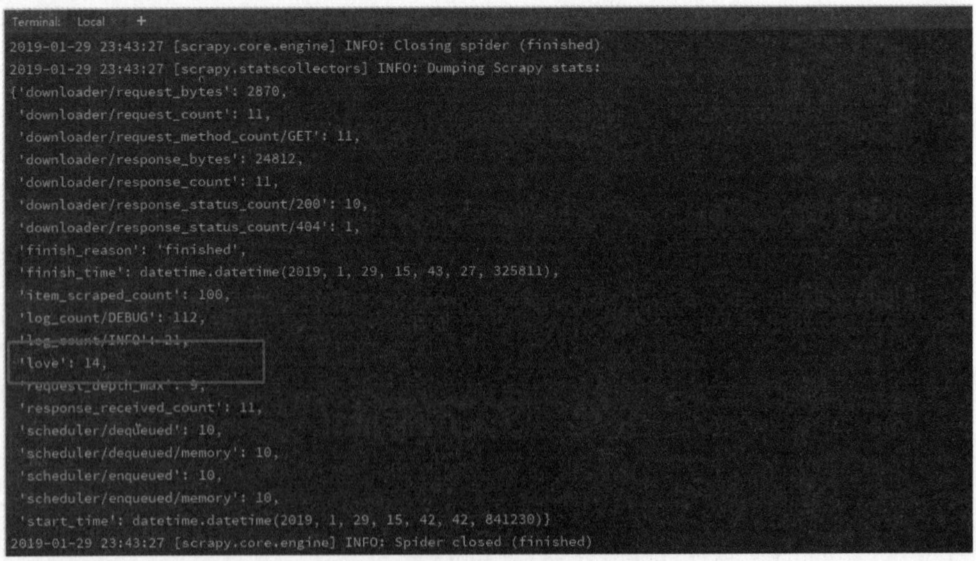

```
Terminal:  Local  +
2019-01-29 23:43:27 [scrapy.core.engine] INFO: Closing spider (finished)
2019-01-29 23:43:27 [scrapy.statscollectors] INFO: Dumping Scrapy stats:
{'downloader/request_bytes': 2870,
 'downloader/request_count': 11,
 'downloader/request_method_count/GET': 11,
 'downloader/response_bytes': 24812,
 'downloader/response_count': 11,
 'downloader/response_status_count/200': 10,
 'downloader/response_status_count/404': 1,
 'finish_reason': 'finished',
 'finish_time': datetime.datetime(2019, 1, 29, 15, 43, 27, 325811),
 'item_scraped_count': 100,
 'log_count/DEBUG': 112,
 'log_count/INFO': 21,
 'love': 14,
 'request_depth_max': 9,
 'response_received_count': 11,
 'scheduler/dequeued': 10,
 'scheduler/dequeued/memory': 10,
 'scheduler/enqueued': 10,
 'scheduler/enqueued/memory': 10,
 'start_time': datetime.datetime(2019, 1, 29, 15, 42, 42, 841230)}
2019-01-29 23:43:27 [scrapy.core.engine] INFO: Spider closed (finished)
```

图 8.3 stats 统计数据

Scrapy 内置可用的数据收集器除了基本的 StatsCollector 外，还有基于 StatsCollector 的其他数据收集器，可以通过 STATS_CLASS 设置来选择。Scrapy 默认使用的是 MemoryStatsCollector，它的原型如下：

```
class scrapy.statscollectors.MemoryStatsCollector
```

这是一个非常简单的数据收集器。其在 Spider 运行完毕后将数据保存在内存中。数据可以通过 spider_stats 属性访问，该属性以字典类型保存了每个 Spider 最近一次爬取的状态的数据。

8.4　发送邮件

虽然 Python 通过 smtplib 库可以很方便地发送 E-Mail，Scrapy 仍然提供了对邮件功能的实现。该功能十分易用，同时由于采用了 Twisted 非阻塞式（non-blocking）IO，其避免了对爬虫的非阻塞式 IO 的影响。另外，Scrapy 也提供了简单的 API 来发送附件。通过一些简单的 settings 设置，可以很方便地发送邮件。

8.4.1　简单例子

有两种方法可以创建邮件发送器（Mail Sender）。可以通过标准构造器（Constructor）创建：

```
from scrapy.mail import MailSender
mailer = MailSender()
```

或者通过传递一个 Scrapy 设置对象，通过 settings 配置创建：

```
mailer = MailSender.from_settings(settings)
```

再使用 send()方法来发送邮件（不包括附件）：

```
mailer.send(to=["someone@example.com"], subject="Some subject", body="Some body", cc=["another@example.com"])
```

8.4.2　MailSender 类

下面来详细介绍 MailSender 类。

原型：

```
class scrapy.mail.MailSender(smtphost=None, mailfrom=None, smtpuser=None, smtppass=None, smtpport=None)
```

参数：

- smtphost (str) ：发送 E-Mail 的 SMTP 主机（host）。如果忽略，就使用 MAIL_HOST。
- mailfrom(str): 用于发送 E-Mail 的地址(address)(填入 From:)。如果忽略，就使用 MAIL_FROM。
- smtpuser: SMTP 用户。如果忽略，就使用 MAIL_USER。如果未给定，将不会进行 SMTP 认证（Authentication）。
- smtppass(str): SMTP 认证的密码。

- smtpport(int)：SMTP 连接的端口。
- smtptls：强制使用 STARTTLS。
- smtpssl(boolean)：强制使用 SSL 连接。

方法：

（1）from_settings(settings)

使用 Scrapy 设置对象来初始化对象。其中，参数 settings(scrapy.settings.Settings 类)可从 settings.py 中设置的值获取。

（2）send(to, subject, body, cc=None, attachs=(), mimetype='text/plain')

发送 E-Mail 到给定的接收者。其参数介绍如下。

- to (list)：字符串或者字符串列表，指定收件人地址。
- subject (str)：邮件主题。
- cc (str or list)：抄送人地址。
- body (str)：邮件正文。
- attachs (iterable)：可迭代的元组，形式为(attach_name, mimetype, file_object)。其中，attach_name 为附件的文件名，mimetype 是附件的 MIME 类型，file_object 是包含附件内容的可读的文件对象。
- mimetype (str)：E-Mail 的 MIME 类型。

8.4.3 在 settings.py 中对 Mail 进行设置

settings.py 中的设置定义了 MailSender 构造器的默认值。

- MAIL_FROM：默认值为 scrapy@localhost，用于发送 E-Mail 的地址（address）（填入 From:）。
- MAIL_HOST：默认值为 localhost，发送 E-Mail 的 SMTP 主机（host）。
- MAIL_PORT：默认值为 25，发用邮件的 SMTP 端口。
- MAIL_USER：默认值为 None，SMTP 用户。如果未给定，将不会进行 SMTP 认证。
- MAIL_PASS：默认值为 None，用于 SMTP 认证，与 MAIL_USER 匹配的认证密码。
- MAIL_TLS：默认值为 False，强制使用 STARTTLS。STARTTLS 能够在已经存在的不安全连接的基础上，通过使用 SSL/TLS 来实现安全连接。
- MAIL_SSL：默认值为 False，强制使用 SSL 加密连接。

8.5 实战：抓取猫眼电影 TOP100 榜单数据

【示例 8-2】抓取猫眼电影 TOP100 榜单数据

抓取的数据为电影名称、主演、上映日期、评分。将抓取的数据保存到 maoyantop100.json 文件，并将文件作为附件通过邮件发送给接收人。

8.5.1　分析页面元素

分析页面元素，如图 8.4 所示。

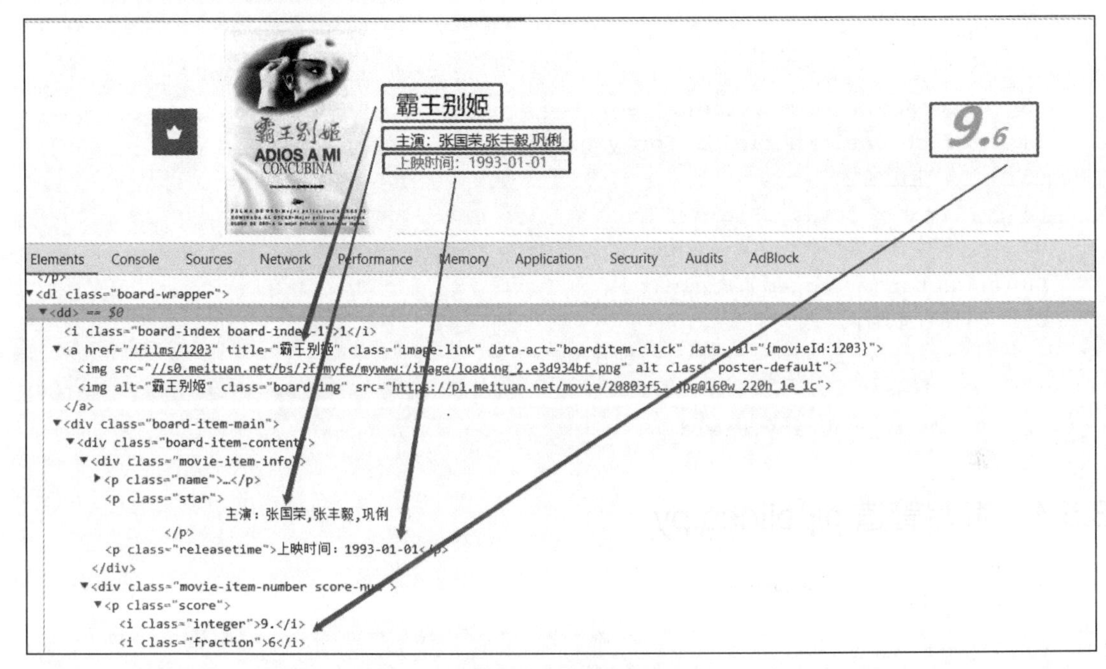

图 8.4　页面结构

可以看到，每一个电影的介绍信息包含在一个<dd>标签中，电影名称在<a>标签的 title 属性中，主演在“class=star”的<p>标签的文本中，上映时间在“class=releasetime”的<p>标签的文本中，评分可以分别在“class=integer”和“class=fraction”的<i>标签中提取数据组合，也可以直接在“class=score”的<p>标签中通过正则表达式提取数据。

8.5.2　创建项目

创建项目命令如下：

```
>>>scrapy startproject maoyan
New Scrapy project 'maoyan', using template ******

You can start your first spider with:
    cd maoyan
scrapy genspider example example.com
>>> scrapy genspider -t crawl top100 maoyan.com
Created spider 'top100' using template 'crawl' in module:
  maoyan.spiders.top100
```

8.5.3 编写 items.py

分析数据提取方法之后，编写代码。其中 items.py 代码如下：

```
01 import scrapy
02
03 class MaoyanItem(scrapy.Item):
04     # define the fields for your item here like:
05     # 电影名称
06     name = scrapy.Field()
07     # 主演
08     actors = scrapy.Field()
09     # 上映时间
10     releasetime = scrapy.Field()
11     # 评分
12     score = scrapy.Field()
```

8.5.4 编写管道 pipelines.py

编写管道 pipelines.py 文件代码：

```
01 import json
02
03 class MaoyanPipeline(object):
04     # 爬虫启动时创建文件 maoyan.json
05     def open_spider(self,spider):
06         self.file = open('maoyantop100.json', 'w')
07
08     # 爬虫关闭时关闭文件
09     def close_spider(self,spider):
10         self.file.close()
11
12     # 将抓取数据写入 JSON 文件
13     def process_item(self, item, spider):
14         line = json.dumps(dict(item),ensure_ascii=False) + "\n"
15         self.file.write(line)
16         return item
```

别忘了在 settings.py 中启用管道，顺便将 ROBOTSTXT_OBEY 值设定为 False：

```
# Obey robots.txt rules
ROBOTSTXT_OBEY = False
***
# Configure item pipelines
```

```
# See https://doc.scrapy.org/en/latest/topics/item-pipeline.html
ITEM_PIPELINES = {
   'maoyan.pipelines.MaoyanPipeline': 300,
}
```

8.5.5　编写爬虫文件 top100.py

编写爬虫文件 top100.py 代码，注意数据提取：

```
01 # -*- coding: utf-8 -*-
02 import scrapy
03 from scrapy.linkextractors import LinkExtractor
04 from scrapy.mail import MailSender
05 from scrapy.spiders import CrawlSpider, Rule
06 from maoyan.items import MaoyanItem
07
08 class TopmoviesSpider(CrawlSpider):
09     name = 'top100'
10     allowed_domains = ['maoyan.com']
11     start_urls = ['https://maoyan.com/board/4']
12
13     # 跟进每页电影目录
14     rules = (
15         Rule(LinkExtractor(allow=r'offset'),callback='parse_item',
               follow=True),
16     )
17
18     def parse_item(self, response):
19         movies = response.css('dd')
20         for movie in movies:
21             item = MaoyanItem()
22             item['name'] = movie.css('a::attr(title)').extract_first()
23             item['actors']=movie.css('.star::text').re_first(r'主演：(.*)')
24             item['releasetime'] = movie.css('.releasetime').
                                         re_first(r'上映时间：(.*)</p>')
25             # 使用正则获取评分的组成部分，如[9,.,7]，其中“9”“7”分别为评分的整数
               与小数部分，其小数点“.”组合之后 9.7 添加到 item 中
26             score = movie.css('.score').re(r'\d|\.')
27             item['score'] = ''.join(score)
28             yield item
29
30     def closed(self,reason):
31         # 使用 settings 中的设置初始化邮件实例
32         mail = MailSender.from_settings(self.settings)
33         # 将需要发送的附件数据使用'rb'模式打开
```

```
34        files = open('./maoyantop100.json', 'rb')
35        # 注意 attachment 是一个迭代器，每一个数据包含 3 部分：
36        # 1.附件的文件名
37        # 2.附件格式
38        # 3.需要发送的附件
39        attachment = [('maoyan_top_100_movies.json','application/json',
                      files)]
40        # 发送邮件，to 指定接收人列表，subject 为邮件主题，body 为邮件正文，
          attachs 为附件，mimetype 为邮件正文类型
41        mail.send(
42           to=['***********@qq.com'],
43           subject=u'maoyan movie',
44           body=u'this is a test',
45           attachs=attachment,
46           mimetype='text/plain',
47           )
48        files.close()
```

在 settings.py 中进行邮件发送配置：

```
# 发送邮件设置
# 指定邮件发送方
MAIL_FROM = 'learnscrapy@163.com'
# 邮件发送服务器
MAIL_HOST = 'smtp.163.com'
# 发送人
MAIL_USER = 'learnscrapy'
# 注意，这是 163 邮箱授权码，不是邮箱密码，不同邮件获取授权码方式不一样
MAIL_PASS = '********'
```

数据提取依据 8.5.1 小节页面元素的分析即可正确提取，在邮件发送部分，我们在 closed()方法中添加邮件发送代码，这部分代码会在爬虫运行结束时执行。

在实例化时，MailSender 的类方法 from_setting(settings)通过 self.settings 获取在 settings.py 中的邮件设置。

在指定附件时，要以'rb'模式打开文件。

运行结果如图 8.5 所示。

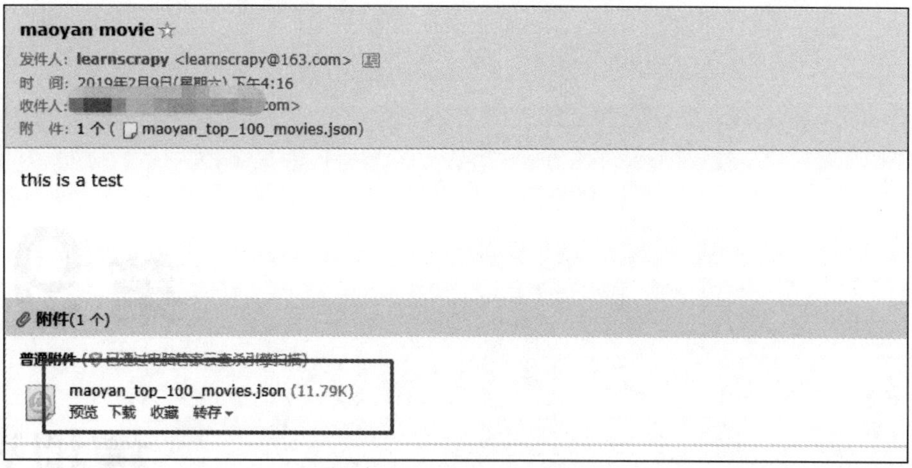

maoyan movie ☆
发件人：learnscrapy <learnscrapy@163.com>
时 间：2019年2月9日(星期六) 下午4:16
收件人：com>
附 件：1 个 (maoyan_top_100_movies.json)

this is a test

附件(1 个)

普通附件 (已通过电脑管家云查杀引擎扫描)

maoyan_top_100_movies.json (11.79K)
预览 下载 收藏 转存

图 8.5 邮箱接收附件结果

第9章 模拟登录

在实际的爬虫作业中，由于一些网站的授权机制，某些内容只能登录后查看，或对非登录限制访问，因此模拟登录在爬虫中经常用到。在使用 Scrapy 处理这种情况时，我们既可以使用表单进行模拟登录，又可以直接使用 Cookie 进行验证，本章将对这两种方法进行介绍。

本章主要的知识点有：

- *表单模拟登录*
- *Cookie 登录*

9.1 模拟提交表单

在 Scrapy 中，模拟提交表单需要用到 Request 的 FormRequest 子类，与 Request 相比，FormRequest 多了一个 formdata 参数，此项参数为要填充的 HTML 表单，是一个字典类型数据，或者可迭代的(key,value)型元组数据。

【示例 9-1】 FormRequest 简单使用方法

```
01 import scrapy
02
03 from myprojct.items import ExampleItem
04
05
06 class ExampleSpider(scrapy.Spider):
07     name = 'example'
08     allowed_domains = ["example.com"]
09
10     start_urls = [
```

```
11          'http://www.example.com',
12      ]
13
14      # 先登录
15      def start_requests(self):
16          return [scrapy.FormRequest(
17              "http://www.example.com/login",
18              # 传递表单数据
19              formdata={'user': 'john', 'pass': 'secret'},
20              # 回调函数
21              callback=self.login_check)]
22
23      # 检查是否登录成功
24      def login_check(self, response):
25
26          # 如果登录成功，则从 start_url 生成 Request，调用 parse_page 进行解析
27          if "Login failed" not in response.body:
28              for url in self.start_urls:
29                  yield scrapy.Request(url, callback=self.parse_page)
30
31      # 解析页面
32      def parse_page(self, response):
33          item = ExampleItem()
34          item["name"] = response.css(".name").extract_first()
35
```

某些网站的登录页面中会有一些含有默认值的隐藏的表单字段包含在<.input type="hidden">元素中，如会话数据、认证信息等。我们在抓取数据时，并不需要了解这些默认数据的生成方法，只需关注需要手动录入数据的字段，如用户名、密码等。这时可以使用 from_response()方法。

from_response()方法原型：

```
    classmethod from_response(response[, formname=None, formid=None,
formnumber=0, formdata=None, formxpath=None, formcss=None, clickdata=None,
dont_click=False, ...])
```

参数说明：

- response：包含待填充数据的 HTML 表单。
- formname：如果指定了 formname 值，那么将使用 name 属性为此值的表单。
- formid：如果指定了 formid 值，那么将使用 id 属性为此值的表单。
- formnumer：如果指定了 formnumber 值，那么将使用序号为此值的表单，第一个表单序号为 0。
- formxpath：如果指定了 formxpath，就使用该 XPath 表达式匹配的第一个 form 表单。
- formcss：如果指定了 formcss，就使用该 CSS 表达式匹配的第一个 form 表单。

- formdata：表单填充数据，如果某个字段在 Response 中已经存在值，那么将会被覆盖；如果某个字段传递的是 None，那么该字段不会包含在生成的 Request 中。
- clickdata：指定表单中的单击事件，如果没有指定，就会通过模拟单击表单中第一个可单击元素进行表单数据提交。
- dont_click：如果指定为 True，该表单就不会通过单击任何元素来操作而直接提交。

from_response()是通过模拟自动单击表单中的可单击元素（如<input type="submit">）来进行表单数据提交的，生成一个 Request。虽然很方便，但仍存在一些问题，例如，如果表单数据是通过 JavaScript 来进行交互操作的，那么 from_response()的默认提交操作就不合适了。这时就可以设置 don’t_click=True 来禁用自动单击提交功能。

【示例 9-2】from_response()简单示例

```
01 import scrapy
02
03 class LoginSpider(scrapy.Spider):
04     name = 'example.com'
05     start_urls = ['http://www.example.com/users/login.php']
06
07     def parse(self, response):
08         return scrapy.FormRequest.from_response(
09             response,
10             # 传递表单数据
11             formdata={'username': 'john', 'password': 'secret'},
12             # 回调函数
13             callback=self.after_login
14         )
15
16     def after_login(self, response):
17         # 检查是否登录成功
18         if "authentication failed" in response.body:
19             self.logger.error("Login failed")
20             return
21         # 执行登录通过后的操作
22
```

【代码解析】

from_response()接收的 Response 来自 start_request()方法从 start_url 列表中加载的 URL 生成的 Response，这个 Response 中包含登录页面默认填充的隐藏表单的数据，因此在 formdata 参数中只需填充登录使用的数据即可，最后调用 after_login()进行登录后续处理。

9.2　用 Cookie 模拟登录状态

在第 2 章介绍 requests 库的时候讲到过 Cookie，Cookie 中包含用户的登录信息等数据。同样的，在 Scrapy 的 FormRequest 中，也可以使用 Cookie 直接登录，进行后续数据的抓取工作。

使用 Cookie 实现登录其实就是把已登录的信息（用户名、密码及其他验证信息）一起发给服务器做验证。优点是不需要知道登录 URL 和表单字段，也不需要了解登录过程和其他细节，即可实现必须登录后查看的目标网页的数据采集。不足之处是 Cookie 有有效期限制，有效期过后，需要重新获取 Cookie 的值。

【示例 9-3】 使用 Cookie 登录简单示例

```
01 import scrapy
02
03 from myprojct.items import ExampleItem
04
05
06 class ExampleSpider(scrapy.Spider):
07     name = 'example'
08     allowed_domains = ["example.com"]
09
10     start_urls = [
11         'http://www.example.com',
12     ]
13
14     # 先登录
15     def start_requests(self):
16         # Cookies 数据
17         cookies = {'uid': '"1083428ut78j8"', 'v': '30'}
18         # 头信息
19         headers = {
20             'Connection': 'keep-alive',
21             'User-Agent': 'Mozilla/5.0 (Windows NT 10.0; Win64; x64)
                            AppleWebKit/537.36 22 (KHTML, like Gecko)
                            Chrome/72.0.3626.121 Safari/537.36'
23         }
24
25         return [scrapy.FormRequest("http://www.example.com/articles",
26                                    # 使用 Cookies
27                                    cookies=cookies,
28                                    # 指定头信息
29                                    headers=headers,
30                                    # 指定回调函数
```

```
31                              callback=self.parse_page)]
32
33
34      # 解析页面
35      def parse_page(self, response):
36          item = ExampleItem()
37          item["name"] = response.css(".name").extract_first()
38          yield item
39
```

在 start_requests 中，FormRequest 直接传入了 cookies 进行登录，添加了头信息，然后调用回调函数处理数据。

9.3 项目实战

下面通过两个实例来进一步讲解表单登录与 Cookie 登录的方法。这两种方法读者都应该掌握，并能根据实际情况灵活运用。

9.3.1 实战 1：使用 FormRequest 模拟登录豆瓣

【示例 9-4】使用 FormRequest 模拟登录豆瓣

（1）确定登录接口

打开豆瓣，进入登录页面，在登录面板选择密码登录，打开浏览器开发者工具，以 FireFox 浏览器为例，切换到网络标签页，选择筛选 XHR 请求，同时勾选“持续日志”复选框，以便在加载页面时保留请求记录，如图 9.1 所示。

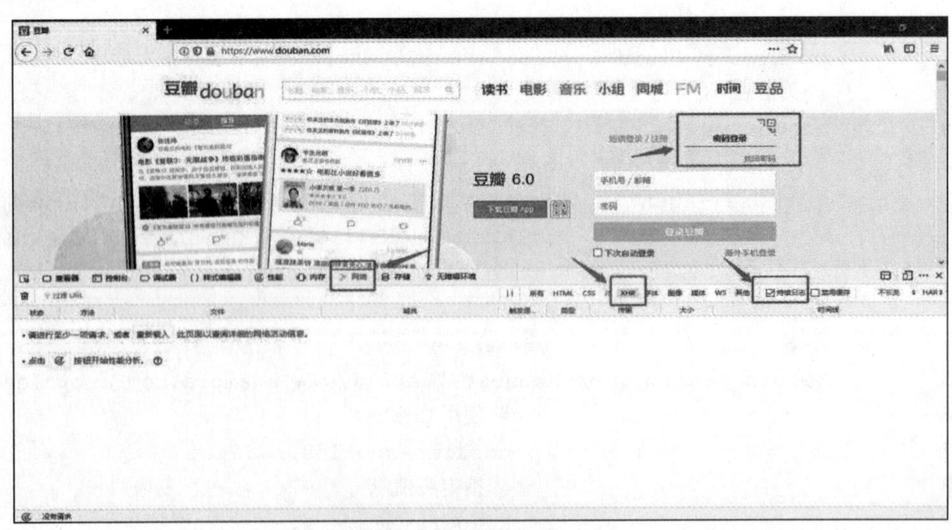

图 9.1　豆瓣登录页面分析

输入用户名和密码进行登录，查看请求记录，如图 9.2 所示。

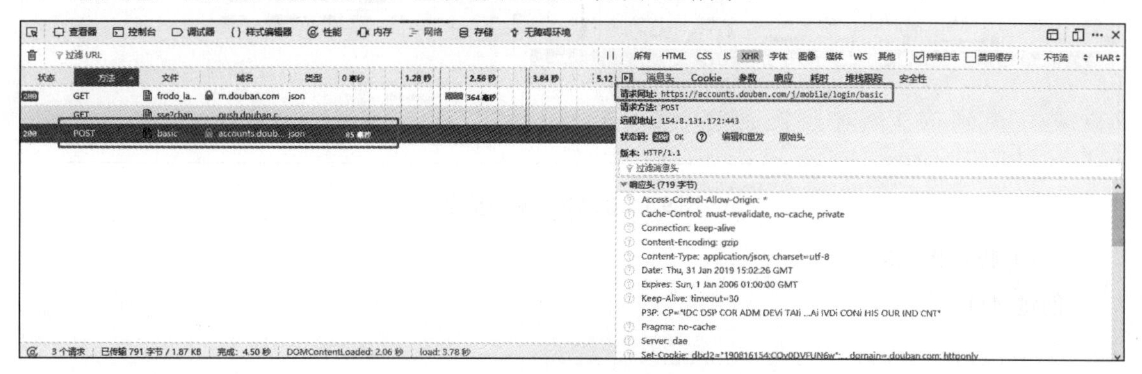

图 9.2　豆瓣登录接口分析

可以看到请求文件为 basic 对应的请求网址 https://accounts.douban.com/j/mobile/login/basic 就是登录的接口，并且消息头标签数据中有响应头与请求头信息，切换到参数标签页，可以看到接口的请求参数如图 9.3 所示。

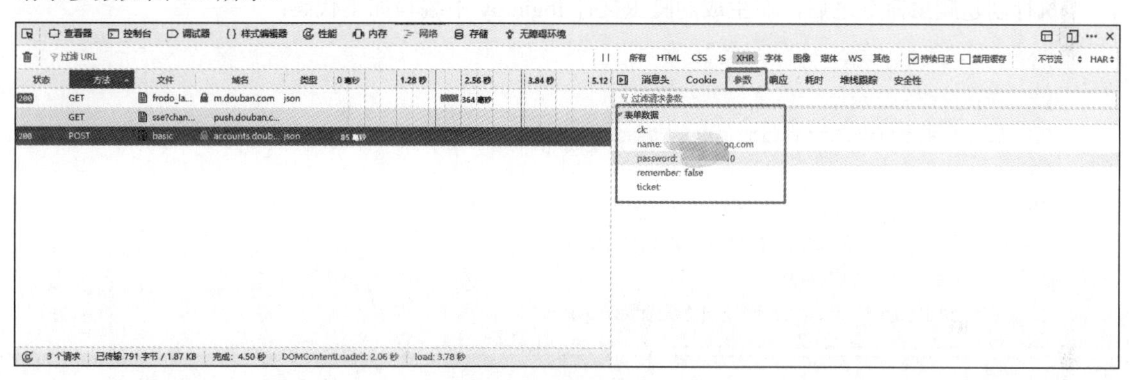

图 9.3　登录参数分析

响应标签页则为接口的响应信息，登录成功响应信息如图 9.4 所示。

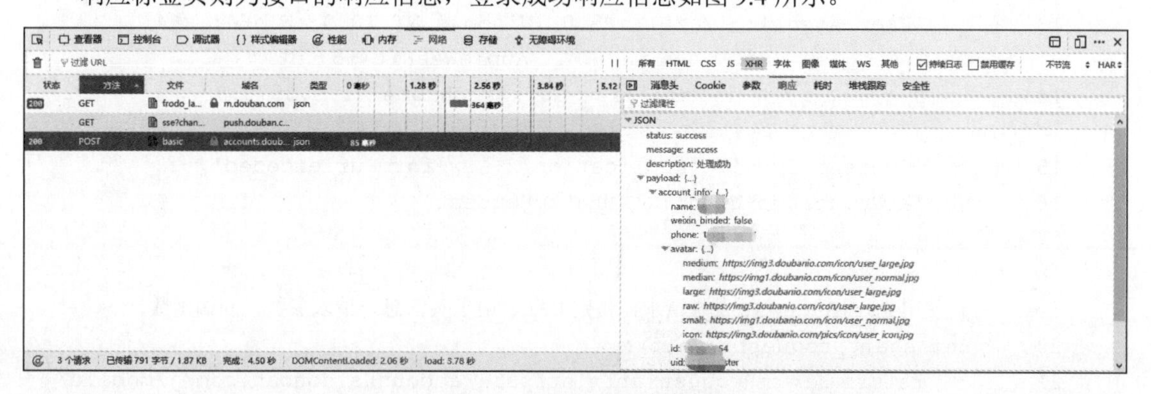

图 9.4　登录成功响应信息

登录失败响应信息如图 9.5 所示。

有这些信息之后，我们就可以使用 Scrapy 的 FormRequest 来模拟提交表单进行登录操作。

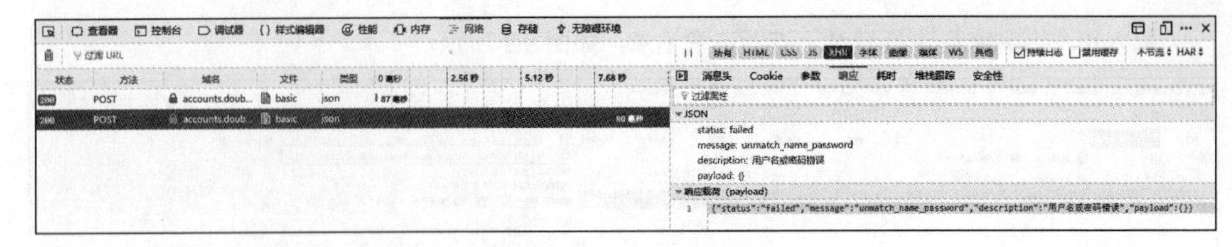

图 9.5　登录失败响应信息

（2）爬虫代码编写

创建项目：

```
>>>scrapy startproject douban
```

创建爬虫：

```
>>>cd douban
>>>scrapy genspider login douban.com
```

执行创建爬虫命令之后，在生成的爬虫文件 login.py 中编写如下代码：

```
01 # -*- coding: utf-8 -*-
02 import scrapy
03 from scrapy.http import FormRequest
04
05
06 class LoginSpider(scrapy.Spider):
07     name = 'login'
08     allowed_domains = ['douban.com']
09     start_urls = ['https://www.douban.com/']
10
11     # 请求头信息，豆瓣会禁止 Scrapy 默认的头信息
12     headers = {
13         'User-Agent': 'Mozilla/5.0 (Windows NT 10.0; Win64; x64)
                                      AppleWebKit/537.36 '
14                       '(KHTML, like Gecko) Chrome/71.0.3578.98
                                      Safari/537.36',
15         'Content-Type': 'application/x-www-form-urlencoded',
16         'X-Requested-With': 'XMLHttpRequest'
17     }
18
19     # 使用 FormRequest 发送请求，指定 URL、请求头信息、请求参数、回调函数
20     def start_requests(self):
21         return [FormRequest(url='https://accounts.douban.com/j/mobile/
                                login/basic',
22                             headers=self.headers,
23                             # 表单数据
24                             formdata={'name': '********',
25                                       'password': '********',
```

```
26                                 'remember': 'false'},
27                         # 回调函数
28                         callback=self.login_check)]
29
30     # 检查登录状态，登录成功后回调爬虫处理函数
31     def login_check(self, response):
32         if "success" in response.body.decode('utf-8'):
33             for url in self.start_urls:
34                 yield scrapy.Request(url=url,
35                                      headers=self.headers,
36                                      callback=self.parse)
37
38     # 爬虫处理函数
39     def parse(self, response):
40         user_check = response.css('.nav-user-account > a > span::text').
                                                    extract_first()
41         self.logger.info('{}已经登录成功'.format(user_check))
42
```

同时，在 settings.py 文件中将 ROBOTSTXT_OBEY 设置为 False。

```
# Obey robots.txt rules
ROBOTSTXT_OBEY = False
```

【代码分析】

在 login.py 文件中，我们添加了浏览器头信息，包含 User-Agent 用户系统信息、Content-Type 资源类型、X-Request-With 请求类型，替换这些默认的头信息之后，服务器可以接收请求信息，豆瓣服务器依据这些信息来判断是不是正常用户请求。

重写 start_request()，这里使用 FormRequest 请求登录网址，传入字典类型请求参数 formdata、headers 头信息，指定后续处理函数 login_check，用来检查是否登录成功。

在 login_check()中，我们在确定登录接口时，知道了返回的信息，因此我们可以用登录失败时响应数据中的“failed”关键字来确定接口是否返回了登录失败的信息。如果登录成功，就请求 start_url 列表中的 url，调用 parse()进行后续处理。

豆瓣登录成功后，会返回首页。在 parse()中，我们在首页中提取出用户名，如图 9.6 所示。

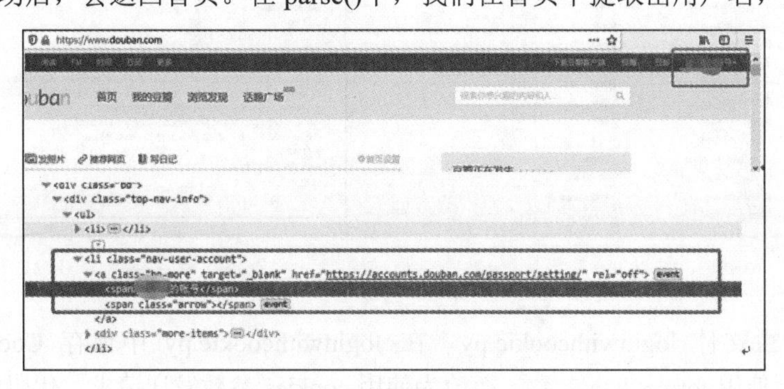

图 9.6　登录后用户名提取

登录失败结果如图 9.7 所示。

```
2019-03-09 22:23:35 [scrapy.core.engine] INFO: Spider opened
2019-03-09 22:23:35 [scrapy.extensions.logstats] INFO: Crawled 0 pages (at 0 pages/min), scraped 0 items (at 0 items/min)
2019-03-09 22:23:35 [scrapy.extensions.telnet] DEBUG: Telnet console listening on 127.0.0.1:6023
2019-03-09 22:23:35 [scrapy.core.engine] DEBUG: Crawled (200) <POST https://accounts.douban.com/j/mobile/login/basic> (referer: None)
2019-03-09 22:23:35 [login] INFO: 用户名或密码错误
2019-03-09 22:23:35 [scrapy.core.engine] INFO: Closing spider (finished)
2019-03-09 22:23:35 [scrapy.statscollectors] INFO: Dumping Scrapy stats:
{'downloader/request_bytes': 461,
 'downloader/request_count': 1,
 'downloader/request_method_count/POST': 1,
 'downloader/response_bytes': 734,
 'downloader/response_count': 1,
 'downloader/response_status_count/200': 1,
```

图 9.7　登录失败结果

登录成功结果如图 9.8 所示。

```
2019-03-09 22:11:02 [scrapy.core.engine] INFO: Spider opened
2019-03-09 22:11:02 [scrapy.extensions.logstats] INFO: Crawled 0 pages (at 0 pages/min), scraped 0 items (at 0 items/min)
2019-03-09 22:11:02 [scrapy.extensions.telnet] DEBUG: Telnet console listening on 127.0.0.1:6023
2019-03-09 22:11:02 [scrapy.core.engine] DEBUG: Crawled (200) <POST https://accounts.douban.com/j/mobile/login/basic> (referer: None)
2019-03-09 22:11:04 [scrapy.core.engine] DEBUG: Crawled (200) <GET https://www.douban.com/> (referer: https://accounts.douban.com/j/mobile/login/basic
2019-03-09 22:11:04 [login] INFO: [illegible]的帐号已经登录成功
2019-03-09 22:11:04 [scrapy.core.engine] INFO: Closing spider (finished)
2019-03-09 22:11:04 [scrapy.statscollectors] INFO: Dumping Scrapy stats:
{'downloader/request_bytes': 952,
 'downloader/request_count': 2,
 'downloader/request_method_count/GET': 1,
 'downloader/request_method_count/POST': 1,
 'downloader/response_bytes': 18913,
 'downloader/response_count': 2,
 'downloader/response_status_count/200': 2,
```

图 9.8　登录成功结果

9.3.2　实战 2：使用 Cookie 登录

回到登录接口的分析中，在登录成功后，我们随便打开一个其他的页面，如豆瓣读书，查看 Cookie 中记录的请求 Cookie 信息，如图 9.9 所示。

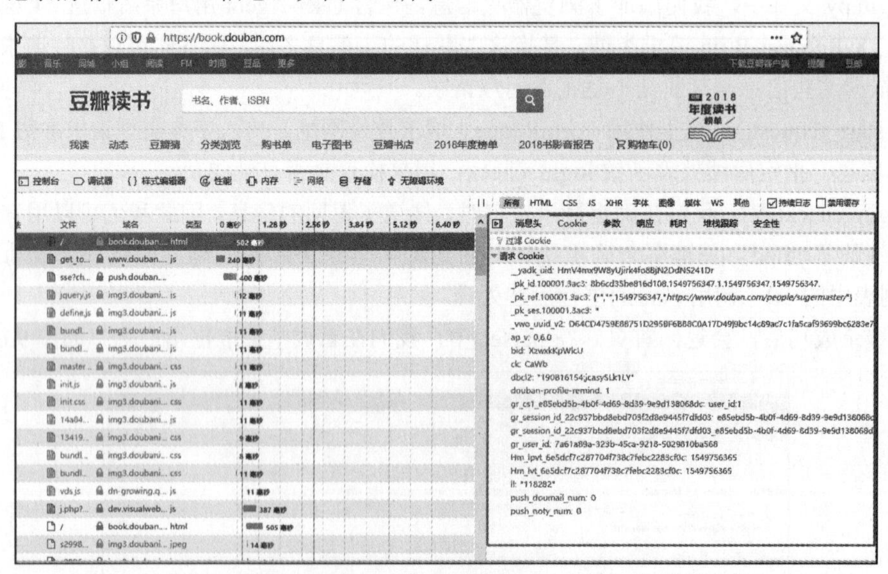

图 9.9　Cookie 分析

新建一个爬虫文件 loginwithcookie.py，在 loginwithcookie.py 中保存 Cookie 信息，修改 start_request()，不使用 formatdata 参数，而改为使用 cookies 参数发送请求，代码如下：

```
01 # -*- coding: utf-8 -*-
02 import scrapy
03 from scrapy import FormRequest
04
05 class LoginwithcookieSpider(scrapy.Spider):
06     name = 'loginwithcookie'
07     allowed_domains = ['douban.com']
08     start_urls = ['http://douban.com/']
09
10     cookies = {
11         '__yadk_uid': 'HmV4mx9W8yUjirk4fo8BjN2DdNS241Dr',
12         '_pk_id.100001.3ac3':
               '8b6cd35be816d108.1549756347.1.1549756347.1549756347.',
13         '_pk_ref.100001.3ac3': '["","",1549756347,
               "https://www.douban.com/people/sugermaster/"]',
14         '_pk_ses.100001.3ac3': '*',
15         '_vwo_uuid_v2': 'D64CD4759E8B751D295BF6BB8C0A17D49|
                                       9bc14c89ac7c1fa5caf93699bc6283e7',
16         'ap_v': '0,6.0',
19         'bid': 'XzwxkKpWlcU',
18         'ck': 'CaWb',
19         'dbcl2': '"190816154:jcasySLk1LY"',
20         'douban-profile-remind': '1',
21         'gr_cs1_e85ebd5b-4b0f-4d69-8d39-9e9d138068dc': 'user_id:1',
22         'gr_session_id_22c937bbd8ebd703f2d8e9445f7dfd03':
                                 'e85ebd5b-4b0f-4d69-8d39-9e9d138068dc',
23
24
25 'gr_session_id_22c937bbd8ebd703f2d8e9445f7dfd03_e85ebd5b-4b0f-4d69-
   8d39-9e9d1380   68dc': 'true',
26
27         'gr_user_id': '7a61a89a-323b-45ca-9218-5029810ba568',
28         'Hm_lpvt_6e5dcf7c287704f738c7febc2283cf0c': '1549756365',
29         'Hm_lvt_6e5dcf7c287704f738c7febc2283cf0c': '1549756365',
30         'll': '"118282"',
31         'push_doumail_num': '0',
32         'push_noty_num': '0'
33     }
34
35     headers = {
36         'Accept': 'text/html,application/xhtml+xm…plication/xml;q=0.9,
                 */*;q=0.8',
37         'Accept-Encoding': 'gzip, deflate, br',
38         'Accept-Language': 'zh-CN,zh;q=0.8,zh-TW;q=0.7,zh-HK;q=0.5,
                             en-US;q=0.3,en;q=0.2',
```

```
39
40         'Connection': 'keep-alive',
41         'DNT': '1',
42         'Host': 'book.douban.com',
43         'Referer': 'https://www.douban.com/people/sugermaster/',
44         'Upgrade-Insecure-Requests': '1',
45         "User-Agent": "Mozilla/5.0 (Windows NT 10.0; Win64; x64; rv': '64.0)
                       Gecko/20100101 Firefox/64.0"
46
47 }
48
49     # 使用 FormRequest 发送请求，指定 url、请求头信息、cookies
50     def start_requests(self):
51         for url in self.start_urls:
52             return [FormRequest(url,
53                                 headers=self.headers,
54                                 # formdata={'name': '1120844583@qq.com',
55                                 #           'password': 'guoqing1010',
56                                 #           'remember': 'false'},
57                                 cookies=self.cookies,
58                                 callback=self.parse)]
59
60     # 爬虫处理函数
61     def parse(self, response):
62         user_check = response.css(
63             '.nav-user-account > a > span::text').extract_first()
64         self.logger.info('{}已经登录成功'.format(user_check))
```

运行爬虫，查看日志记录，如图 9.10 所示。

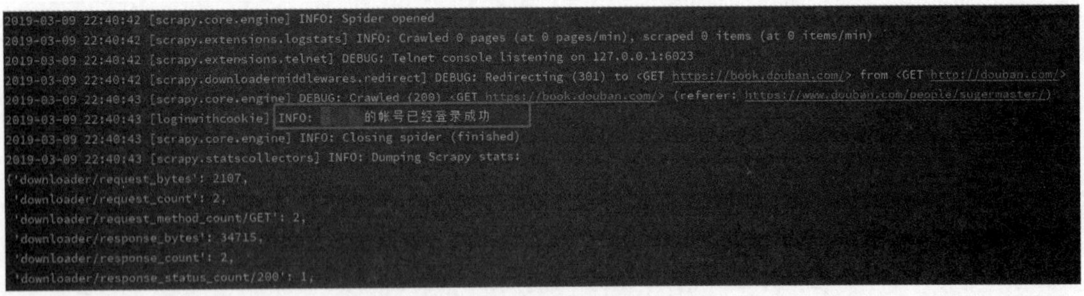

```
2019-03-09 22:40:42 [scrapy.core.engine] INFO: Spider opened
2019-03-09 22:40:42 [scrapy.extensions.logstats] INFO: Crawled 0 pages (at 0 pages/min), scraped 0 items (at 0 items/min)
2019-03-09 22:40:42 [scrapy.extensions.telnet] DEBUG: Telnet console listening on 127.0.0.1:6023
2019-03-09 22:40:42 [scrapy.downloadermiddlewares.redirect] DEBUG: Redirecting (301) to <GET https://book.douban.com/> from <GET http://douban.com/>
2019-03-09 22:40:43 [scrapy.core.engine] DEBUG: Crawled (200) <GET https://book.douban.com/> (referer: https://www.douban.com/people/sugermaster/)
2019-03-09 22:40:43 [loginwithcookie] INFO:       的帐号已经登录成功
2019-03-09 22:40:43 [scrapy.core.engine] INFO: Closing spider (finished)
2019-03-09 22:40:43 [scrapy.statscollectors] INFO: Dumping Scrapy stats:
{'downloader/request_bytes': 2107,
 'downloader/request_count': 2,
 'downloader/request_method_count/GET': 2,
 'downloader/response_bytes': 34715,
 'downloader/response_count': 2,
 'downloader/response_status_count/200': 1,
```

图 9.10 使用 Cookie 登录成功结果

第 10 章

Scrapy 爬虫优化

通过前面章节介绍的 Scrapy 使用方法，我们已经可以进行大部分网站的爬取工作。但我们仍然需要思考一些问题：怎样使爬虫达到最优的下载速度？如何避免被目标网站加入黑名单、被禁止访问？如何避免重复抓取。因此，在爬虫可以正常工作后，下一步需要进行的工作就是进行爬虫项目的优化。

本章主要的知识点有：

- Scrapy 性能评估
- Scrapy 性能优化

10.1 Scrapy+MongoDB 实战：抓取并保存 IT 之家博客新闻

下面我们将通过一个爬取示例来演示如何一步一步优化爬虫。

【示例 10-1】使用 Scrapy+MongoDB 抓取并保存 IT 之家（www.ithome.com）博客新闻

10.1.1 确定目标

在 IT 之家首页打开一篇新闻，我们需要抓取的数据有本章标题、文章地址、发布日期、来源、原文章地址、作者、文章标签，如图 10.1 所示。

小米：今年西欧零售店数量将增长2倍

2019-2-26 20:14:55 来源：新浪科技 作者：李明 责编：浮生 评论：12

北京时间2月26日晚间消息，全球第五大智能手机厂商小米公司今日向CNBC表示，到今年年底，小米在西欧的零售店数量将增长2倍。

2017年，小米在西班牙开设了其在欧洲的首家零售店。今日，小米集团高级副总裁王翔在世界移动通信大会(MWC)上接受CNBC采访时称，到2017年底，**希望小米在西欧的零售店数量能超过150家。**

而截至2018年底，小米在西欧的零售店数量**不到50家**。王翔说："这对我们来说是一个大目标。"

与此同时，中国另一家手机厂商OPPO本月早些时候曾表示，将进军英国等三个欧洲新市场。

市场研究公司IDC数据显示，去年第四季度OPPO超越小米，成为全球第四大智能机厂商。而前三大厂商分别为三星电子、苹果公司和华为。

文章价值 打分人数少于5人，暂不显示得分　无价值 0　还可以 0　有价值 1

分享：   　下载IT之家APP，分享赚金币换豪礼

相关文章　关键词：小米，小米之家

图 10.1　IT 之家博客新闻

10.1.2　创建项目

先创建项目：

```
>>>scrapy startproject ithome
New Scrapy project 'ithome'…
```

创建爬虫：

```
>>>cd ithome
>>>scrapy genspider -t crawl news ithome.com
Created spider 'news' using template 'crawl' in module:
  ithome.spiders.news
```

10.1.3 编写 items.py 文件

在 10.1.1 节，我们确定了需要抓取的内容，现在开始编写 items.py，代码如下：

```
01 # -*- coding: utf-8 -*-
02
03 # Define here the models for your scraped items
04 #
05 # See documentation in:
06 # https://doc.scrapy.org/en/latest/topics/items.html
07
08 import scrapy
09
10
11 class IthomeItem(scrapy.Item):
12     # define the fields for your item here like:
13
14     # 文章标题
15     title = scrapy.Field()
16     # 文章 URL
17     url = scrapy.Field()
18     # 来源
19     source = scrapy.Field()
20     # 来源 URL
21     source_url = scrapy.Field()
22     # 发布日期
23     release_date = scrapy.Field()
24     # 作者
25     author = scrapy.Field()
26     # 关键词
27     key_words = scrapy.Field()
```

定义了需要抓取的内容：文章标题、文章 URL、来源、来源 URL、发布日期、作者、关键词。

10.1.4 编写爬虫文件 news.py

这里我们使用 Crawl 模板创建爬虫，这是因为每一篇新闻中都包含其他的新闻链接，利用 rule 参数可以方便地跟踪新闻页面，抓取数据。

待抓取元素的定位不再详细介绍。这里推荐一个插件 ChroPath，FireFox 与 Chrome 都可以安装，可以显示元素的绝对与相对 XPath、CSS 选择器，如图 10.2 所示。

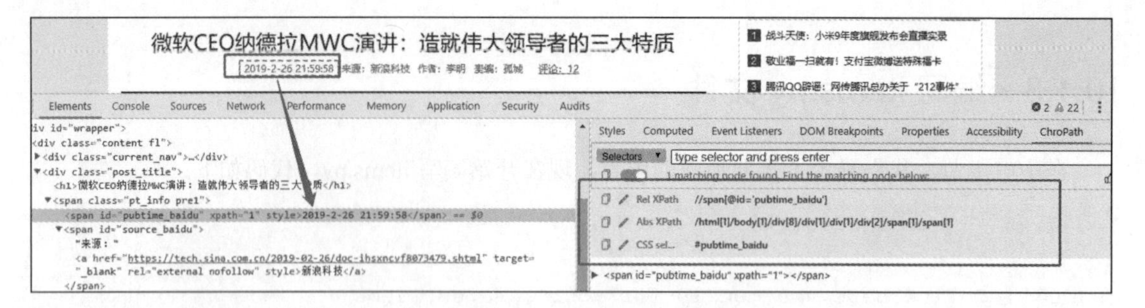

图 10.2 ChroPath 插件

new.py 代码如下：

```
01 # -*- coding: utf-8 -*-
02 import scrapy
03 from scrapy.linkextractors import LinkExtractor
04 from scrapy.spiders import CrawlSpider, Rule
05 from ithome.items import IthomeItem
06 import datetime
07
08
09 class NewsSpider(CrawlSpider):
10     name = 'news'
11     allowed_domains = ['ithome.com']
12     start_urls = ['https://www.ithome.com/0/411/151.htm']
13
14     rules = (
15         # 文章 URL 形如 https://www.ithome.com/0/411/369.htm
16         # 根据后三段数字来提取所有的文章 url 并跟进处理数据
17         Rule(LinkExtractor(allow=r'/\d/\d{3}/\d{3}'),
                              callback='parse_item', follow=True),
18     )
19
20     def parse_item(self, response):
21         item = IthomeItem()
22         # 文章 URL
23         item['url'] = response.url
24         # 文章标题
25         item['title'] = response.css('.post_title > h1::text').
                                                     extract_first()
26         # 文章作者
27         item['author'] = response.css('#author_baidu strong::text').
                                                     extract_first()
28         # 文章来源
29         item['source'] = response.css('#source_baidu > a::text').
                                                     extract_first()
30         # 文章来源 URL
```

```
31         item['source_url'] = response.css('#source_baidu > a::attr(href)').
                                            extract_first()
32         # 发布日期
33         item['release_date'] = response.css('#pubtime_baidu::text').
                                               extract_first()
34         # 关键词
35         item['key_words'] = response.css('.hot_tags > span a::text').
                                               extract()
36         return item
37
38     def close(spider, reason):
39         self.crawler.stats.set_value('finish_time',datetime.datetime.now())
40
```

文章的URL从Response的url属性中提取即可。一个好的习惯是，我们在做元素提取时，始终在Scrapy Shell中运行检测后再写入代码中，确保不会因为元素提取问题引出异常。代码中重写了close()方法，在爬虫关闭时记录关闭时间，写入数据统计参数finish_time中。

10.1.5 编写管道pipelines.py

在管道中如前面的例子一样，将数据保存在MongoDB中，pipelines.py代码如下：

```
01 # -*- coding: utf-8 -*-
02
03 # Define your item pipelines here
04 #
05 # Don't forget to add your pipeline to the ITEM_PIPELINES setting
06 # See: https://doc.scrapy.org/en/latest/topics/item-pipeline.html
07 from pymongo import MongoClient
08 from scrapy.exceptions import DropItem
09 import datetime
10
11 class IthomePipeline(object):
12     # 定义集合ithome_news
13     collection = 'ithome_news'
14
15     def __init__(self,mongo_uri,mongo_db,stats):
16         self.mongo_uri = mongo_uri
17         self.mongo_db = mongo_db
18         self.stats = stats
19
20     @classmethod
21     def from_crawler(cls,crawler):
22         return cls(
23             # 从settings.py中获取MONGODB数据库连接信息，数据统计参数
```

```
24            mongo_uri=crawler.settings.get('MONGO_URI'),
25            mongo_db=crawler.settings.get('MONGO_DB'),
26            stats=crawler.stats,
27        )
28
29    # 爬虫启动时打开数据库
30    def open_spider(self,spider):
31        self.client = MongoClient(self.mongo_uri)
32        self.db = self.client[self.mongo_db]
33        # 开启爬虫时将结束时间写入数据集参数 start 中
34        self.stats.set_value('start', datetime.datetime.now())
35
36    # 爬虫关闭时关闭数据库连接
37    def close_spider(self,spider):
38        self.client.close()
39
40    def process_item(self, item, spider):
41        # 如果抓取得 item 中含有 title，则为有效数据，保存，否则丢弃
42        if not item['title']:
43            raise DropItem("数据不完整，丢弃：{}".format(item))
44        else:
45            self.db[self.collection].insert_one(dict(item))
46        return item
47
```

我们在爬虫开启时添加了一个统计数据参数 start，用于记录爬虫的开启时间。默认的统计参数中有 start_time，只不过此参数记录的开始时间是第一个 url 下载以后开始处理数据的时间。

10.1.6 编写 settings.py

在 settings.py 文件中，我们添加 MongoDB 数据库的连接信息、用 Pipeline，另外，还需要设定一个 CLOSESPIDER_ITEMCOUNT 参数，在抓取到一定数量的 Item 后，关闭爬虫，这里我们定位 10 000 条数据。settings.py 添加、修改代码如下：

```
# Configure item pipelines
# See https://doc.scrapy.org/en/latest/topics/item-pipeline.html
ITEM_PIPELINES = {
   'ithome.pipelines.IthomePipeline': 300,
}

# 当爬取到 10000 条数据时关闭爬虫
CLOSESPIDER_ITEMCOUNT = 10000

# MongoDB
MONGO_URI = 'localhost:27017'
MONGO_DB = 'ithome'
```

10.1.7　运行爬虫

代码编写完毕后，运行爬虫，查看运行结果，如图 10.3 所示。

```
'downloader/request_count': 10178,
'downloader/request_method_count/GET': 10178,
'downloader/response_bytes': 77061666,
'downloader/response_count': 10178,
'downloader/response_status_count/200': 10062,
'downloader/response_status_count/403': 115,
'downloader/response_status_count/404': 1,
'dupefilter/filtered': 61679,
'finish_reason': 'closespider_itemcount',
'finish_time': datetime.datetime(2019, 3, 10, 23, 1, 48, 414580),
'httperror/response_ignored_count': 115,
'httperror/response_ignored_status_count/403': 114,
'httperror/response_ignored_status_count/404': 1,
'item_dropped_count': 40,
'item_dropped_reasons_count/DropItem': 40,
'item_scraped_count': 10020,
'log_count/DEBUG': 20200,
'log_count/INFO': 128,
'log_count/WARNING': 40,
'request_depth_max': 156,
'response_received_count': 10178,
'scheduler/dequeued': 10176,
'scheduler/dequeued/memory': 10176,
'scheduler/enqueued': 12737,
'scheduler/enqueued/memory': 12737,
'start': datetime.datetime(2019, 3, 10, 22, 55, 21, 948759),
```

图 10.3　爬取结果 1

从图 10.3 中可以看到，共发送了 10178 次请求，成功 200 状态 10062 条，其中异常状态数据：403（Forbidden，服务器理解请求客户端的请求，但是拒绝执行此请求）状态 115 条，404（Not found，服务器无法根据客户端的请求找到资源）状态 1 条，爬虫开始时间 22：55：21，结束时间 23：01：48，得知爬虫运行了 6 分钟左右，爬虫的运行时间能否缩短，以提升速度？需要注意的是，在日志中，那些为 403 状态的数据依然可以打开，如图 10.4 所示，是什么原因导致这种问题，如何进行避免，也就是爬虫质量如何提升？

```
2019-02-27 22:31:36 [scrapy.core.engine] DEBUG: Crawled (403) <GET https://www.ithome.com/0/386/916.htm> (referer: https://www.ithome.com/0/387/862.htm)
2019-02-27 22:31:36 [scrapy.core.engine] DEBUG: Crawled (403) <GET https://www.ithome.com/0/388/798.htm> (referer: https://www.ithome.com/0/389/323.htm)
2019-02-27 22:31:36 [scrapy.core.scraper] DEBUG: Scraped from <200 https://www.ithome.com/0/388/553.htm>
{'author': '过客',
 'key_words': ['沙特', 'Uber', '投资', '安全'],
 'release_date': '2018-10-14 17:10:07',
 'source': '网易科技',
 'source_url': 'http://tech.163.com/18/1014/08/DU2L68LC00097U7R.html',
 'title': 'Uber CEO等商业大佬拒绝到沙特参加投资会议：不安全，怕出事',
 'url': 'https://www.ithome.com/0/388/553.htm'}
2019-02-27 22:31:36 [scrapy.spidermiddlewares.httperror] INFO: Ignoring response <403 https://www.ithome.com/0/389/056.htm>: HTTP status code is not handled or not allowed
2019-02-27 22:31:36 [scrapy.core.scraper] DEBUG: Scraped from <200 https://www.ithome.com/0/388/388.htm>
{'author': '李丽',
 'key_words': ['Facebook', '微信'],
 'release_date': '2018-10-13 9:19:37',
 'source': '新浪科技',
 'source_url': 'https://tech.sina.com.cn/i/2018-10-13/doc-ifxeuwws3743143.shtml',
 'title': 'Facebook Messenger终于要支持消息撤回了',
 'url': 'https://www.ithome.com/0/388/388.htm'}
2019-02-27 22:31:36 [scrapy.core.engine] DEBUG: Crawled (200) <GET https://www.ithome.com/0/387/315.htm> (referer: https://www.ithome.com/0/388/126.htm)
2019-02-27 22:31:36 [scrapy.spidermiddlewares.httperror] INFO: Ignoring response <403 https://www.ithome.com/0/386/916.htm>: HTTP status code is not handled or not allowed
2019-02-27 22:31:36 [scrapy.spidermiddlewares.httperror] INFO: Ignoring response <403 https://www.ithome.com/0/388/798.htm>: HTTP status code is not handled or not allowed
2019-02-27 22:31:36 [scrapy.core.scraper] DEBUG: Scraped from <200 https://www.ithome.com/0/387/315.htm>
```

图 10.4　爬取结果 2

我们将通过对爬虫的优化设置来解决上述问题。

10.2 用 Benchmark 进行本地环境评估

Benchmark 是 Scrapy 自带的一个简单的性能测试工具，它会在本地创建一个 HTTP 服务器，并以最大可能的速度对其进行爬取。目的是测试本地执行环境的效率，以此获得一个用于对比的基线。这个性能测试使用一个很简单的 Spider，仅仅是跟进连接，并不做其他处理。

执行命令：

```
scrapy bench
```

运行输出：

```
$ scrapy bench
2019-02-28 11:03:09 [scrapy.utils.log] INFO: Scrapy 1.5.1 started (bot: scrapybot)
2019-02-28 11:03:09 [scrapy.utils.log] INFO: Versions: lxml 4.2.5.0, libxml2 2.9.8, cssselect 1.0.3, parsel 1.5.1, w3lib 1.19.0, Twisted 18.9.0, Python 3.7.1 (default, Dec 10 2018, 22:54:23) [MSC v.1915 64 bit (AMD64)], pyOpenSSL 18.0.0 (OpenSSL 1.1.1a  20 Nov 2018), cryptography 2.4.2, Platform Windows-10-10.0.17763-SP0
2019-02-28 11:03:11 [scrapy.crawler] INFO: Overridden settings: {'CLOSESPIDER_TIMEOUT': 10, 'LOGSTATS_INTERVAL': 1, 'LOG_LEVEL': 'INFO'}
2019-02-28 11:03:12 [scrapy.middleware] INFO: Enabled extensions:
['scrapy.extensions.corestats.CoreStats',
 'scrapy.extensions.telnet.TelnetConsole',
 'scrapy.extensions.closespider.CloseSpider',
 'scrapy.extensions.logstats.LogStats']
2019-02-28 11:03:13 [scrapy.middleware] INFO: Enabled downloader middlewares:
['scrapy.downloadermiddlewares.httpauth.HttpAuthMiddleware',
 'scrapy.downloadermiddlewares.downloadtimeout.DownloadTimeoutMiddleware',
 'scrapy.downloadermiddlewares.defaultheaders.DefaultHeadersMiddleware',
 'scrapy.downloadermiddlewares.useragent.UserAgentMiddleware',
 'scrapy.downloadermiddlewares.retry.RetryMiddleware',
 'scrapy.downloadermiddlewares.redirect.MetaRefreshMiddleware',
 'scrapy.downloadermiddlewares.httpcompression.HttpCompressionMiddleware',
 'scrapy.downloadermiddlewares.redirect.RedirectMiddleware',
 'scrapy.downloadermiddlewares.cookies.CookiesMiddleware',
 'scrapy.downloadermiddlewares.httpproxy.HttpProxyMiddleware',
 'scrapy.downloadermiddlewares.stats.DownloaderStats']
2019-02-28 11:03:13 [scrapy.middleware] INFO: Enabled spider middlewares:
['scrapy.spidermiddlewares.httperror.HttpErrorMiddleware',
 'scrapy.spidermiddlewares.offsite.OffsiteMiddleware',
 'scrapy.spidermiddlewares.referer.RefererMiddleware',
 'scrapy.spidermiddlewares.urllength.UrlLengthMiddleware',
```

```
    'scrapy.spidermiddlewares.depth.DepthMiddleware']
    2019-02-28 11:03:13 [scrapy.middleware] INFO: Enabled item pipelines:
    []
    2019-02-28 11:03:13 [scrapy.core.engine] INFO: Spider opened
    2019-02-28 11:03:13 [scrapy.extensions.logstats] INFO: Crawled 0 pages (at
0 pages/min), scraped 0 items (at 0 items/min)
    2019-02-28 11:03:14 [scrapy.extensions.logstats] INFO: Crawled 53 pages (at
3180 pages/min), scraped 0 items (at 0 items/min)
    2019-02-28 11:03:15 [scrapy.extensions.logstats] INFO: Crawled 117 pages (at
3840 pages/min), scraped 0 items (at 0 items/min)
    2019-02-28 11:03:16 [scrapy.extensions.logstats] INFO: Crawled 181 pages (at
3840 pages/min), scraped 0 items (at 0 items/min)
    2019-02-28 11:03:17 [scrapy.extensions.logstats] INFO: Crawled 237 pages (at
3360 pages/min), scraped 0 items (at 0 items/min)
    2019-02-28 11:03:18 [scrapy.extensions.logstats] INFO: Crawled 293 pages (at
3360 pages/min), scraped 0 items (at 0 items/min)
    2019-02-28 11:03:19 [scrapy.extensions.logstats] INFO: Crawled 341 pages (at
2880 pages/min), scraped 0 items (at 0 items/min)
    2019-02-28 11:03:20 [scrapy.extensions.logstats] INFO: Crawled 389 pages (at
2880 pages/min), scraped 0 items (at 0 items/min)
    2019-02-28 11:03:21 [scrapy.extensions.logstats] INFO: Crawled 437 pages (at
2880 pages/min), scraped 0 items (at 0 items/min)
    2019-02-28 11:03:22 [scrapy.extensions.logstats] INFO: Crawled 485 pages (at
2880 pages/min), scraped 0 items (at 0 items/min)
    2019-02-28 11:03:23 [scrapy.core.engine] INFO: Closing spider
(closespider_timeout)
    2019-02-28 11:03:23 [scrapy.extensions.logstats] INFO: Crawled 525 pages (at
2400 pages/min), scraped 0 items (at 0 items/min)
    2019-02-28 11:03:24 [scrapy.statscollectors] INFO: Dumping Scrapy stats:
    {'downloader/request_bytes': 234184,
     'downloader/request_count': 541,
     'downloader/request_method_count/GET': 541,
     'downloader/response_bytes': 1592745,
     'downloader/response_count': 541,
     'downloader/response_status_count/200': 541,
     'finish_reason': 'closespider_timeout',
     'finish_time': datetime.datetime(2019, 2, 28, 3, 3, 24, 603556),
     'log_count/INFO': 17,
     'request_depth_max': 19,
     'response_received_count': 541,
     'scheduler/dequeued': 541,
     'scheduler/dequeued/memory': 541,
     'scheduler/enqueued': 10819,
     'scheduler/enqueued/memory': 10819,
     'start_time': datetime.datetime(2019, 2, 28, 3, 3, 13, 674927)}
```

```
2019-02-28 11:03:24 [scrapy.core.engine] INFO: Spider closed
(closespider_timeout)
```

说明本地环境上，Scrapy 能以几乎 2400 页面/分钟的速度抓取数据。然而这只是简单的跟进连接，实际情况下爬虫会做更多的数据处理工作，速度并不可能达到这个测试结果。

10.3 扩展爬虫

10.3.1 增大并发

并发是指 Scrapy 同时处理的 Request 请求的数量。类型有全局限制和局部（针对每个网站）限制。

Scrapy 通过在 settings.py 中的 CONCURRENT_REQUESTS 的设定值来确定请求并发数。然而多数情况下，Scrapy 默认的全局并发限制对高并发请求并不适用，我们需要重新设置这个值。一般情况下需要增大设置值，增加多少取决于设计的爬虫能占用多少 CPU。一般开始可以设置为 100 左右，不过最好的方式是做一些测试，获取 Scrapy 进程并发数与 CPU 使用率之间的关系，理想情况下，应该选择一个能使 CPU 使用率在 80%~90%的并发数。

在 setting.py 中设置：

```
CONCURRENT_REQUESTS = 100
```

10.3.2 关闭 Cookie

前面的章节中，我们知道 Cookie 包含网站的一些认证信息，不幸的是，网站也可以通过 Cookie 记录来禁止爬取，Cookie 还会影响 CPU 使用率。因此，除非真的需要，否则应禁止 Cookie，以提高性能。

在 settings.py 中设置：

```
COOKIES_ENABLED = False
```

10.3.3 关闭重试

对大量的网页数据进行爬取工作时，由于各种不可控的原因，一些网页爬取的失败时有发生。默认情况下，Scrapy 会对失败的 HTTP 请求进行重试以获取数据。但这种操作会减慢爬取的效率，特别是当网站响应很慢时，访问这样的网站会造成超时并且会重试多次。这不仅影响效率，同时占用了爬虫爬取其他网站点的资源。

在 settings.py 中设置：

```
RETRY_ENABLED = False
```

10.3.4 减少下载超时时间

当爬取一个响应很慢的网站时，在设置的超时时间结束前，Scrapy 会持续等待。超时时间太长会占用不必要的资源，减少下载超时能让卡住的连接被尽快放弃，释放资源去处理其他的爬取工作。

在 settings.py 中设置：

```
DOWNLOAD_TIMEOUT = 5
```

10.3.5 关闭重定向

除非对跟进重定向感兴趣，否则应考虑关闭重定向。当进行通用爬取时，一般的做法是保存重定向的地址，并在之后的爬取中进行解析。这保证了每批爬取的 Request 数目在一定的数量，否则重定向循环可能会导致爬虫在某个站点耗费过多资源。

在 settings.py 中设置：

```
REDIRECT_ENABLED = False
```

10.3.6 AutoThrottle 扩展

AutoThrottle 扩展基于 Scrapy 服务器和正在抓取的网站的负载来自动优化爬取速度的扩展，通常称为自动限速扩展，这一扩展设计的目的是：

（1）相比较默认的下载延迟设置来说，自动限速对站点更友好。

（2）能够自动调整 Scrapy 达到最佳的爬取速度，所以使用者无须自己调整下载延迟，只需要定义允许最大并发的请求数，剩下的就由该扩展组件自动完成。

在 Scrapy 中，下载延迟是通过计算建立 TCP 连接到接收到 HTTP 头信息（header）之间的时间来测量的。

先来看看前面介绍的下载延迟设定，设定一个固定的 DOWNLOAD_DELAY 值，再根据 CONCURRENT_REQUESTS_PER_DOMAIN 或者 CONCURRENT_REQUESTS_PER_IP 的启用情况来确定适用对象。假设 DOWNLOAD_DELAY=3，CONCURRENT_REQUESTS_PER_DOMAIN=16，就会每 3/16 秒发送一个请求，以达到 16 并发的要求。但我们知道，由于下载延迟很短，偶尔会有突发的请求，而通常情况下非 200 的响应（比如服务器错误 500）会比带有数据响应的 200 更快地返回给客户端，由于设定了固定并发数，这样会有更多的并发请求会提交到服务器，因此可能会造成更多的错误返回，因为错误可能就是由于高频率的请求导致的。

再来看看 AutoThrottle 自动限速。爬虫总是以 AUTOTHROTTLE_START_DELAY 设定值作为下载延迟开始工作。当响应返回时，得到响应延迟 latency，这时下载延迟就被设定为 latency/N，N 是由 AUTOTHROTTLE_TARGET_CONCURRENCY 设定的。

下一个请求的下载延迟为前一个请求的下载延迟与第二步计算出来的下载延迟的平均值。非 200 的响应延迟并不会通过第二步的计算降低下载延迟，最终调整的下载延迟不会比

DOWNLOAD_DELAY 设定值低，也不会比 AUTOTHROTTLE_MAX_DELAY 设定值高。

下面介绍用来设定 AutoThrottle 扩展的一些设定值。

- AUTOTHROTTLE_ENABLED：默认为 False，表示是否启用 AUTOTHROTTLE 扩展。
- AUTOTHROTTLE_START_DELAY：默认为 5.0，初始化下载延迟。
- AUTOTHROTTLE_MAX_DELAY：默认为 60，最高下载延迟。
- AUTOTHROTTLE_TARGET_CONCURRENCY：默认为 1.0，Scrapy 同时请求目标网站的平均请求数。通过设定该参数，AutoThrottle 可以动态调整请求数，若该值设置得高，则并发请求数增加；若该值设置得小，则请求并发数减少，对目标网站影响更小。

即便打开了 AutoThrottle 扩展，CONCURRENT_REQUESTS_PER_DOMAIN 和 CONCURRENT_REQUESTS_PER_IP 依然生效，也就是说，如果 AUTOTHROTTLE_TARGET_CONCURRENCY 设定值大于 CONCURRENT_REQUESTS_PER_DOMAIN 和 CONCURRENT_REQUESTS_PER_IP 设定值，那么该设定值将不会产生作用。

第 11 章

Scrapy 项目实战：爬取某社区用户详情

关于 Scrapy 的基本知识点已经介绍完毕，下面通过一个完整的爬虫实战项目将各个知识点串联起来。本项目主要抓取 SegmentFault（思否）社区用户基本信息及用户的问答得票数等信息。下面将会对整个项目进行详细讲解。

本章主要的知识点有：

- Cookie 登录
- Scrapy 代码中启动

11.1 项目分析

任何项目的第一步都需要进行项目分析。在这一步我们确定需求（抓取哪些数据），确定抓取策略（是否需要登录、如何保存等）。因此，项目分析是整个项目的基础。

11.1.1 页面分析

照例开始对目标页面进行元素分析，在思否社区随意打开一个用户主页，查看用户信息构成，如图 11.1 所示。

我们将个人主页分成 4 部分，分别进行编号：

- 第一部分，包含用户的个人基本信息，如声望、毕业院校、公司等，称为属性面板。
- 第二部分，用户的行为数据统计，如粉丝人数、关注数、提问数、回答数等，称为统计面板。
- 第三部分，用户的产出数据，包括文章、回答、提问等，称为产品面板。
- 第四部分，主要是用户的技能标签，称为技能面板。

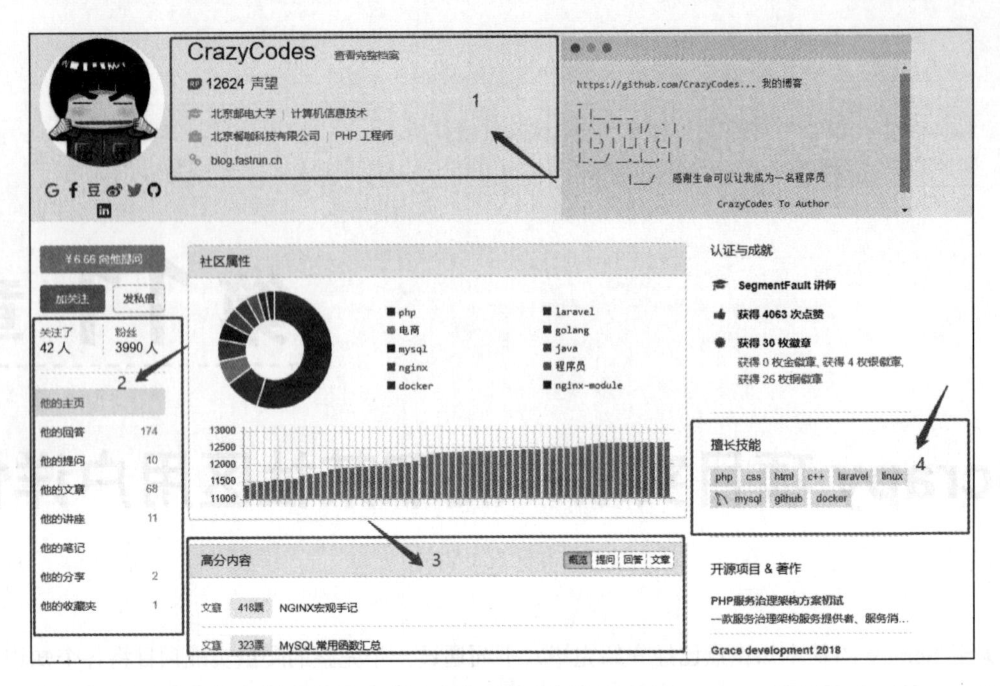

图 11.1 Segmentfault 个人主页

第一部分，属性面板。具体包括用户昵称、声望、毕业院校、专业、公司、职位、个人博客，如图 11.2 所示。

图 11.2　用户属性模块

第二部分，统计面板。我们需要的具体包含关注数、粉丝数、回答数、文章数、讲座数、徽章数，如图 11.3 所示。单击“关注了”人数与“粉丝”人数，可以查看相关联的用户信息，如图 11.4 所示，由此可以进入用户主页，继续抓取用户信息。由于单击“关注了”链接进行用户数据查看时，会跳转到登录页面，因此在爬虫运行时，我们使用 Cookies 进行登录处理。

另一点是需要进行具体徽章数的爬取，单击获取徽章数，会跳转到获得的徽章列表页面，如图 11.5 所示。

第三部分，产品面板。这里我们只关注回答标签页中得票最高的回答，然后提取这个具体问题的内容、问题的类型以及该用户的答案，如图 11.6 所示。

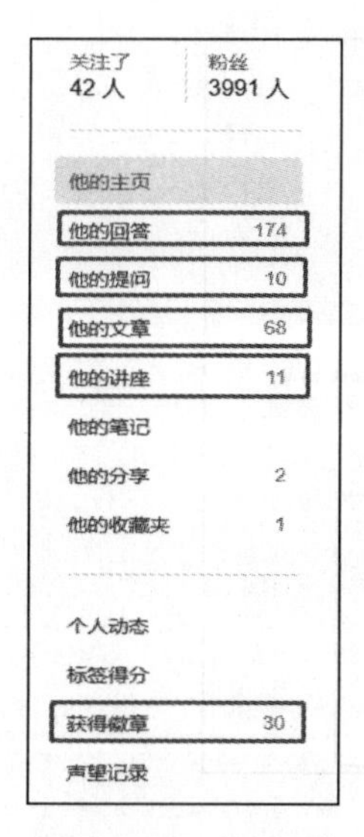

图 11.3　统计面板

关注的标签　关注的人　关注的专栏

TheLarkInn
1037 声望

member
5566 声望

前端小智
13007 声望

xx19941215
1950 声望

Panda
3122 声望

鲔鱼
2721 声望

韩天峰
5547 声望

高阳Sunny
4163 声望

图 11.4　关注的人列表

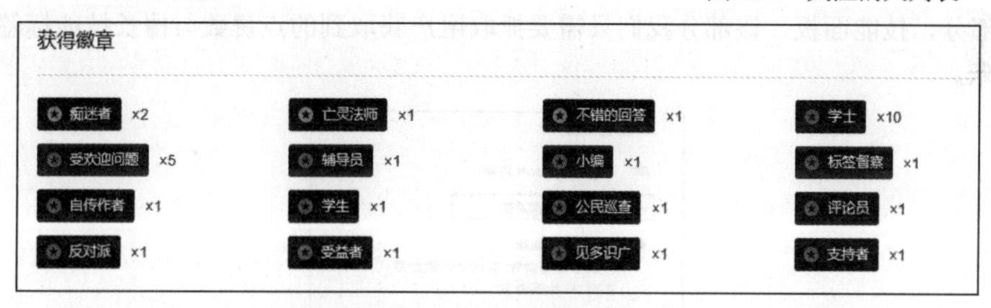

图 11.5　徽章数据页

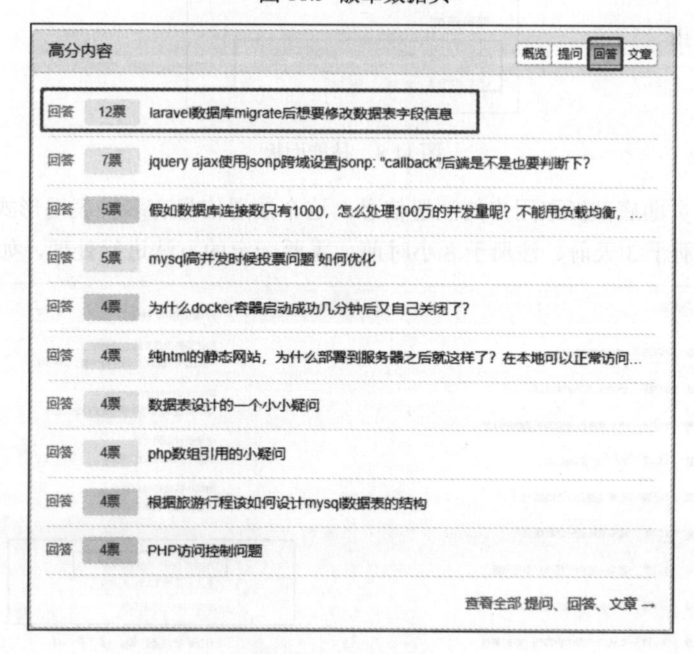

图 11.6　回答列表

得票最高的答案对应问题的信息如图 11.7 所示。

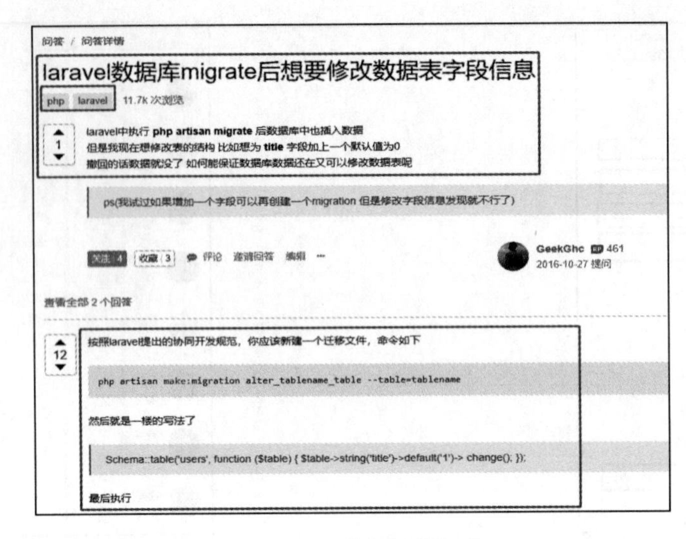

图 11.7　回答问题详细页

第四部分，技能面板。该部分我们只需要抓取用户获取到的点赞数与擅长技能标签即可，如图 11.8 所示。

图 11.8　技能面板

最后，我们在页面底部抓取用户的注册信息。这个注册信息有多种表现形式，如注册于 2018 年 10 月 12 日、注册于 3 天前、注册于 8 小时前，需要一定的方法进行处理，如图 11.9 所示。

图 11.9　注册日期

11.1.2 抓取流程

从页面分析完元素之后，接下来整理数据的抓取流程。

（1）使用 Cookie 进行登录，确保抓取时不会被重定向到登录页面。

（2）从给定的第一个用户主页开始，抓取用户信息。

（3）用户属性信息，统计信息中的数量值信息、技能信息与注册日期信息可以在用户主页进行提取。

（4）徽章信息和回答问题信息需要进入响应的页面进行提取。

（5）保存用户数据到 MongoDB。

（6）从关注人和粉丝列表中提取用户主页链接，保存至调度器。

（7）重复第（3）~（6）步。

用户信息抓取流程图如图 11.10 所示。

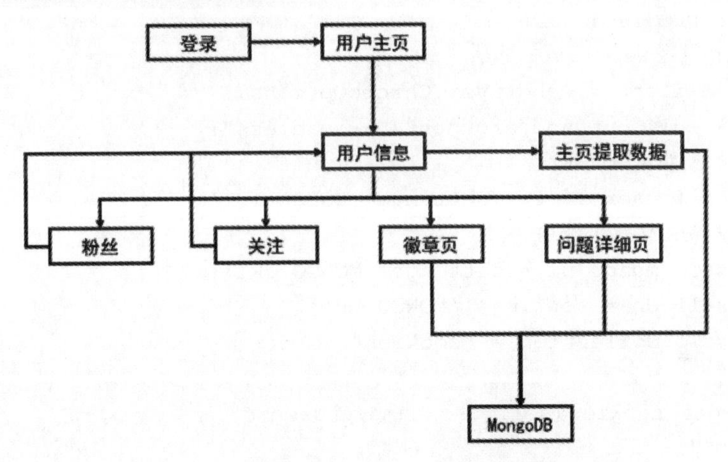

图 11.10 Segmentfault 用户信息抓取流程图

11.2 创建爬虫

分析完项目之后，开始进行具体代码的编写。由于项目中需要用到 Cookie 进行登录，因此我们先要准备一些 cookies。我们可以使用 Selenium 编写一个小程序来收集 cookies。然后，在爬取过程中通过下载器中间件来添加 cookies，通过 Spider 中间件 Response，比如前面介绍的日期的处理。

先来创建项目：

```
>>>scrapy startproject segmentfault
```

创建爬虫时，使用 CrawlerSpider 模板创建：

```
>>>cd segmentfault
>>>scrapy genspider -t crawl userinfo segmentfault.com
```

接下来开始编写部分代码。

11.2.1 cookies收集器

我们使用Selenium来收集cookies。首先我们需要准备几个不同的登录账号，使用不同的账号登录来收集cookies，然后将cookies保存在MongoDB中。在settings.py的同级目录中创建get_cookies.py文件，代码如下：

```
01 from selenium import webdriver
02 from pymongo import MongoClient
03 from scrapy.conf import settings
04
05
06 class GetCookies(object):
07     def __init__(self):
08         # 初始化组件
09         # 设定webdriver选项
10         self.opt = webdriver.ChromeOptions()
11         # self.opt.add_argument("--headless")
12         # 初始化用户列表
13         self.user_list = settings['USER_LIST']
14         # 初始化MongoDB参数
15         self.mongo_ui = settings['MONGO_URI']
16         self.db = settings['MONGO_DB']
17         self.collection = "cookies"
18
19     def get_cookies(self,username,password):
20         """
21         :param username:
22         :param password:
23         :return: cookies
24         """
25         # 使用webdriver选项创建driver
26         driver = webdriver.Chrome(options=self.opt)
27         driver.get("https://segmentfault.com/user/login")
28         driver.find_element_by_name("username").send_keys(username)
29         driver.find_element_by_name("password").send_keys(password)
30         driver.find_element_by_xpath("//button[@type= 'submit']").click()
31         # 登录之后获取页面cookies
32         cookies = driver.get_cookies()
33
34         return cookies
35
36     def format_cookies(self,cookies):
```

```
37          """
38          :param cookies:
39          从 driver.get_cookies 的形式为:
40          [{'domain':'segmentfault.com','httpOnly':False,'name':'PHPSESSID',
41          'path':'/','secure':False,'value':
                                  'web2~5grmfa89j12eksub8hja3bvaq4'},
42          {'domain':'.segmentfault.com','expiry':1581602940,'httpOnly':
                                                                  False,
43          'name': 'Hm_lvt_e23800c454aa573c0ccb16b52665ac26', 'path': '/',
                                                       'secure': False,
44          'value': '1550066940'},
45          {'domain': '.segmentfault.com', 'httpOnly': False,
46          'name': 'Hm_lpvt_e23800c454aa573c0ccb16b52665ac26',
47          'path': '/', 'secure': False, 'value': '1550066940'},
48          {'domain': '.segmentfault.com', 'expiry': 1550067000,
                                                       'httpOnly': False,
49          'name': '_gat', 'path': '/', 'secure': False, 'value': '1'},
50          {'domain': '.segmentfault.com', 'expiry': 1550153340,
                                                       'httpOnly': False,
51          'name': '_gid', 'path': '/', 'secure': False, 'value':
                                      'GA1.2.783265084.1550066940'},
52          {'domain': '.segmentfault.com', 'expiry': 1613138940,
                                  'httpOnly': False, 'name': '_ga',
53          'path': '/', 'secure': False, 'value':
                                  'GA1.2.1119166665.1550066940'}]
54          只需提取每一项的 name 与 value 即可
55
56          :return:
57          """
58          c = dict()
59          for item in cookies:
60              c[item['name']] = item['value']
61
62          return c
63
64      def save(self):
65          client = MongoClient(self.mongo_ui)
66          db = client[self.db]
67          # 从用户列表中获取用户名与密码，分别登录获取 cookies
68          for username,password in self.user_list:
69              cookies = self.get_cookies(username,password)
70              f_cookies = self.format_cookies(cookies)
71              print("insert cookie:{}".format(f_cookies))
72              # 将格式整理后的 cookies 插入 MongoDB 数据库
73              db[self.collection].insert_one(f_cookies)
```

```
74
75 if __name__ == '__main__':
76     cookies = GetCookies()
77     cookies.save()
78
```

同时，settings.py 中的 MongoDB 与用户列表参数设置如下：

```
# 配置 MONGODB
MONGO_URI = 'localhost:27017'
MONGO_DB = 'segmentfault'

# 用户列表
USER_LIST = [
   ("username1","password1"),
   ("username2","password2"),
]
```

get_cookies.py 中代码没有什么需要特别注意的地方，都是前面章节知识点的整合。唯一的技巧是，我们在 settings.py 文件中以(username,password)元组作为 USER_LIST 的元素，这样可以直接遍历 username 与 password 作为 get_cookies()的参数。

查看收集到的 cookies，如图 11.11 所示。

```
> db.cookies.find().pretty()
{
        "_id" : ObjectId("5c642200fed7e7153cea7f2a"),
        "PHPSESSID" : "web1~0jinss4li78pq5h6a9c0pkb0cj",
        "Hm_lvt_e23800c454aa573c0ccb16b52665ac26" : "1550066176",
        "Hm_lpvt_e23800c454aa573c0ccb16b52665ac26" : "1550066176",
        "_gat" : "1",
        "_gid" : "GA1.2.99770019.1550066176",
        "_ga" : "GA1.2.1865447365.1550066176"
}
{
        "_id" : ObjectId("5c642204fed7e7153cea7f2c"),
        "PHPSESSID" : "web2~i6gsn71ut0tjadokae80m80rfg",
        "Hm_lvt_e23800c454aa573c0ccb16b52665ac26" : "1550066179",
        "Hm_lpvt_e23800c454aa573c0ccb16b52665ac26" : "1550066179",
        "_gat" : "1",
        "_gid" : "GA1.2.1572629857.1550066180",
        "_ga" : "GA1.2.270479626.1550066180"
}
{
        "_id" : ObjectId("5c6424c9fed7e7183488fa61"),
        "PHPSESSID" : "web2~2h5ae0rf4oquqk5tmnjp7dl9q6",
        "Hm_lvt_e23800c454aa573c0ccb16b52665ac26" : "1550066889",
        "Hm_lpvt_e23800c454aa573c0ccb16b52665ac26" : "1550066889",
        "_gat" : "1",
        "_gid" : "GA1.2.246154849.1550066889",
        "_ga" : "GA1.2.441106865.1550066889"
}
{
        "_id" : ObjectId("5c6424d0fed7e7183488fa62"),
        "PHPSESSID" : "web2~iirf1id12duk6k126aa235u00i",
        "Hm_lvt_e23800c454aa573c0ccb16b52665ac26" : "1550066895",
        "Hm_lpvt_e23800c454aa573c0ccb16b52665ac26" : "1550066895",
        "_gat" : "1",
        "_gid" : "GA1.2.504334720.1550066895",
        "_ga" : "GA1.2.1273794664.1550066895"
}
```

图 11.11　cookies 收集

11.2.2　Items 类

接着根据我们前面的分析定义 Item，确定数据结构，items.py 代码如下：

```
01 # -*- coding: utf-8 -*-
02
03 # Define here the models for your scraped items
04 #
05 # See documentation in:
06 # https://doc.scrapy.org/en/latest/topics/items.html
07
08 import scrapy
09
10
11 class SegmentfaultItem(scrapy.Item):
12     # define the fields for your item here like:
13     # 个人属性
14     # 姓名
15     name = scrapy.Field()
16     # 声望
17     rank = scrapy.Field()
18     # 学校
19     school = scrapy.Field()
20     # 专业
21     majors = scrapy.Field()
22     # 公司
23     company = scrapy.Field()
24     # 工作
25     job = scrapy.Field()
26     # blog
27     blog = scrapy.Field()
28     # 社交活动数据
29     # 关注人数
30     following = scrapy.Field()
31     # 粉丝数
32     fans = scrapy.Field()
33     # 回答数
34     answers = scrapy.Field()
35     # 提问数
36     questions = scrapy.Field()
37     # 文章数
38     articles = scrapy.Field()
39     # 讲座数
40     lives = scrapy.Field()
```

```
41      # 徽章数
42      badges = scrapy.Field()
43      # 技能属性
44      # 点赞数
45      like = scrapy.Field()
46      # 技能
47      skills = scrapy.Field()
48      # 注册日期
49      register_date = scrapy.Field()
50      # 问答统计
51      # 回答最高得票数
52      answers_top_score = scrapy.Field()
53      # 得票数最高的回答对应的问题的标题
54      answers_top_title = scrapy.Field()
55      # 得票数最高的回答对应的问题的标签
56      answers_top_tags = scrapy.Field()
57      # 得票数最高的回答对应的问题的内容
58      answers_top_question = scrapy.Field()
59      # 得票数最高的回答对应的问题的内容
60      answers_top_content = scrapy.Field()
```

11.2.3 Pipeline 管道编写

本项目中，我们将数据保存至MongoDB，保存方法与前面章节一样，pipelines.py代码如下：

```
01 request_with_cookies = Request(url="http://www.example.com",
02 # -*- coding: utf-8 -*-
03 
04 # Define your item pipelines here
05 #
06 # Don't forget to add your pipeline to the ITEM_PIPELINES setting
07 # See: https://doc.scrapy.org/en/latest/topics/item-pipeline.html
08 import pymongo
09 
10 class SegmentfaultPipeline(object):
11     # 设定MongoDB集合名称
12     collection_name = 'userinfo'
13 
14     def __init__(self,mongo_uri,mongo_db):
15         self.mongo_uri = mongo_uri
16         self.mongo_db = mongo_db
17 
18     # 通过crawler获取settings.py中设定的MongoDB连接信息
19     @classmethod
20     def from_crawler(cls,crawler):
```

```
21         return cls(
22             mongo_uri = crawler.settings.get('MONGO_URI'),
23             mongo_db = crawler.settings.get('MONGO_DB','segmentfault')
24         )
25
26     # 当爬虫启动时连接 MongoDB
27     def open_spider(self,spider):
28         self.client = pymongo.MongoClient(self.mongo_uri)
29         self.db = self.client[self.mongo_db]
30
31     # 当爬虫关闭时断开 MongoDB 连接
32     def close_spider(self,spider):
33         self.client.close()
34
35     # 将 Item 插入数据库保存
36     def process_item(self, item, spider):
37         self.db[self.collection_name].insert_one(dict(item))
38         return item
39
```

不要忘记在 settings.py 中激活编写的 Pipeline：

```
# Configure item pipelines
# See https://doc.scrapy.org/en/latest/topics/item-pipeline.html
ITEM_PIPELINES = {
   'segmentfault.pipelines.SegmentfaultPipeline': 300,
}
```

11.2.4 Spider 爬虫文件

由于我们是使用 Cookie 进行用户验证的，因此需要在 start_request()中进行 Cookie 登录操作，然后保存 Cookie 并传递。登录之后，我们进入初始页面解析用户数据，并跟踪 Rule 中定义的链接继续抓取。

在爬虫文件 userinfo.py 中重写 start_request()，代码如下：

```
01 # -*- coding: utf-8 -*-
02 import scrapy
03 import time
04 from scrapy import Request
05 from pymongo import MongoClient
06 from scrapy.linkextractors import LinkExtractor
07 from scrapy.spiders import CrawlSpider,Rule
08 from scrapy.http import FormRequest
09 from segmentfault.items import SegmentfaultItem
10
```

```
11
12 class UserinfoSpider(CrawlSpider):
13     name = 'userinfo'
14     allowed_domains = ['segmentfault.com']
15     start_urls = ['https://segmentfault.com/u/mybigbigcat/users/
                                 following']
16
17     rules = (
18         # 用户主页地址，跟进并进行解析
19         Rule(LinkExtractor(allow=r'/u/\w+$'),callback='parse_item',
               follow=True),
20         # 用户关注列表，跟进列表页面，抓取用户主页地址进行后续操作
21         Rule(LinkExtractor(allow=r'/users/followed$'),follow=True),
22         # 用户粉丝列表，跟进列表页面，抓取用户主页地址进行后续操作
23         Rule(LinkExtractor(allow=r'/users/following$'),follow=True),
24         # 跟进其他页面地址
25         Rule(LinkExtractor(allow=r'/users/[followed|following]?
               page=\d+'),follow=True),
26     )
27
28     def start_requests(self):
29         # 从 MongoDB 中获取一条 Cookie，添加到开始方法
30         client = MongoClient(self.crawler.settings['MONGO_URI'])
31         db = client[self.crawler.settings['MONGO_DB']]
32         cookies_collection = db.cookies
33         # 获取一条 Cookie
34         cookies = cookies_collection.find_one()
35         # Cookie 中的'Hm_lpvt_e23800c454aa573c0ccb16b52665ac26'
             参数是当前时间的
36          #10 位表示法，因此重新填充
37         cookies['Hm_lpvt_e23800c454aa573c0ccb16b52665ac26'] =
                  str(int(time.time()))
38
39         return [Request("https://segmentfault.com",
40                         cookies=cookies,
41                         callback=self.after_login)]
42
43     # 登录之后从 start_url 中开始抓取数据
44     def after_login(self,response):
45         for url in self.start_urls:
46             return self.make_requests_from_url(url)
47
```

在跟进规则 rules 中，我们需要跟进 4 种链接：

（1）Rule(LinkExtractor(allow=r'/u/\w+$'), callback='parse_item', follow=True)

用户主页链接，如图 11.12 所示。

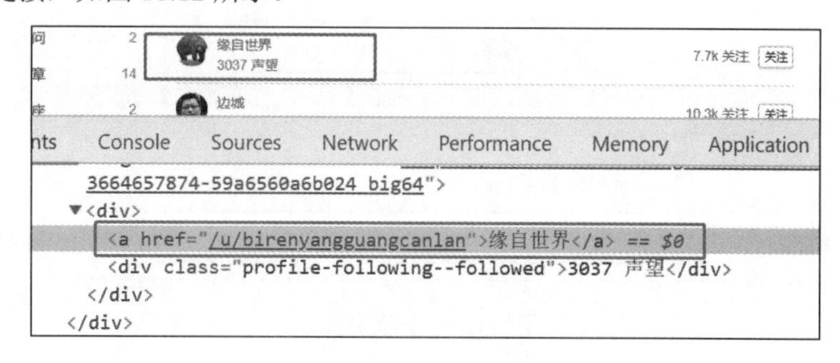

图 11.12　用户主页链接

（2）Rule(LinkExtractor(allow=r'/users/followed$'), follow=True)

用户关注列表页，如图 11.13 所示，记录了该用户关注的用户列表，只需跟进此链接进入列表页即可。

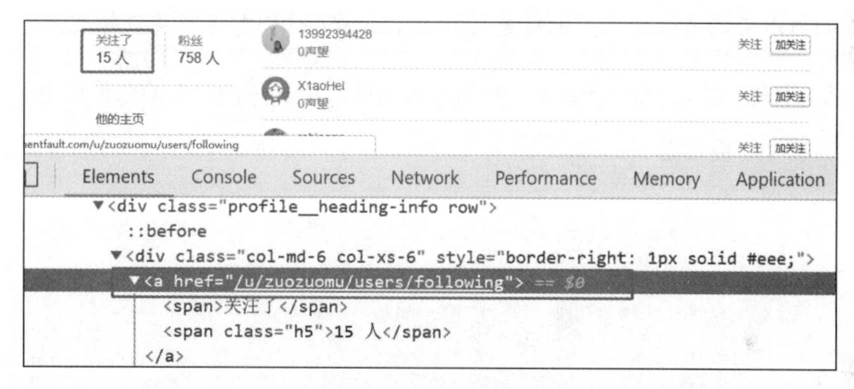

图 11.13　关注用户列表链接

（3）Rule(LinkExtractor(allow=r'/users/following$'), follow=True)

用户粉丝列表页，如图 11.4 所示，记录了该用户的粉丝列表，只需跟进链接进入列表页即可。

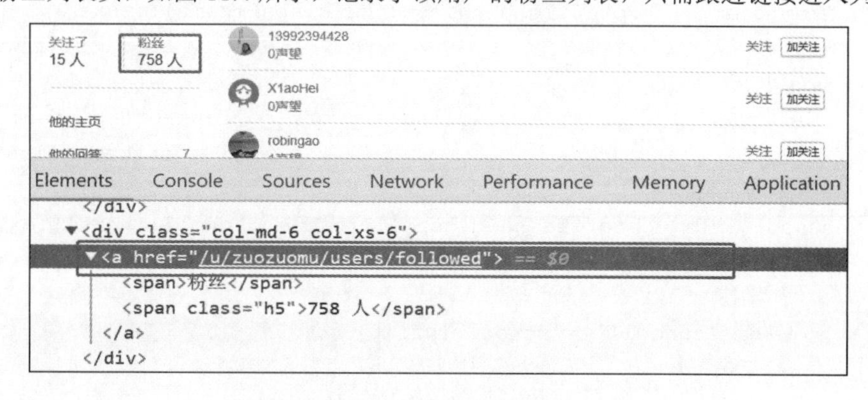

图 11.14　粉丝列表链接

（4）Rule(LinkExtractor(allow=r'/users/[followed|following]?page=\d+'), follow=True)

用户的关注列表或粉丝列表有分页，如图 11.15 所示，跟进每一页的链接。

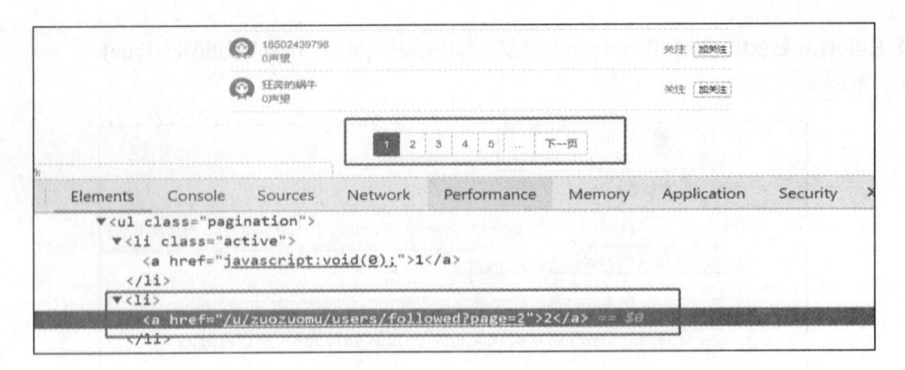

图 11.15　列表分页

在 after_login()方法中，我们使用了 make_request_from_url()方法来处理 start_url 中的链接，这时因为 make_request_from_url()会调用 parse()进行后续处理，这样就能与 CrawSpider 的 parse()关联起来。在 4.2.1 节中我们讲过，CrawSpider 中的 parse()有特殊的作用不能进行复写，但当重写 start_request()时仍需将其关联起来。

用户信息提取的方法为 parse_item()，在用户主页中，可以直接提取的数据如下：

（1）个人属性面板数据

```
48 def parse_item(self, response):
49     """
50     :param response:
51     :return:request with item as meta
52     """
53     item = SegmentfaultItem()
54     # 个人属性
55     profile_head = response.css('.profile__heading')
56     # 姓名
57     item['name'] = profile_head.css('h2[class*=name]::text').
                                              re_first(r'\w+')
58     # 声望
59     item['rank'] = profile_head.css('.profile__rank-btn > span::text').
                                              extract_first()
60     # 学校专业信息
61     school_info = profile_head.css('.profile__school::text').extract()
62     if school_info:
63         # 学校
64         item['school'] = school_info[0]
65         # 专业
66         item['majors'] = school_info[1].strip()
67     else:
68         item['school'] = ''
69         item['majors'] = ''
```

```
70      # 公司职位信息
71      company_info = profile_head.css('.profile__company::text').extract()
72      if company_info:
73          # 公司
74          item['company'] = company_info[0]
75          # 职位
76          item['job'] = company_info[1].strip()
77      else:
78          item['company'] = ''
79          item['job'] = ''
80      # 个人博客
81      item['blog'] = profile_head.css('a[class*=other-item-link]::
                                    attr(href)').extract_first()
82
```

提取元素时，要注意灵活选择方法，包含换行、空格等非打印字符时，首选正则表达式。在处理学校及公司时，我们需要加一点判断。先看一下学校与公司的元素定位，如图 11.16 所示。

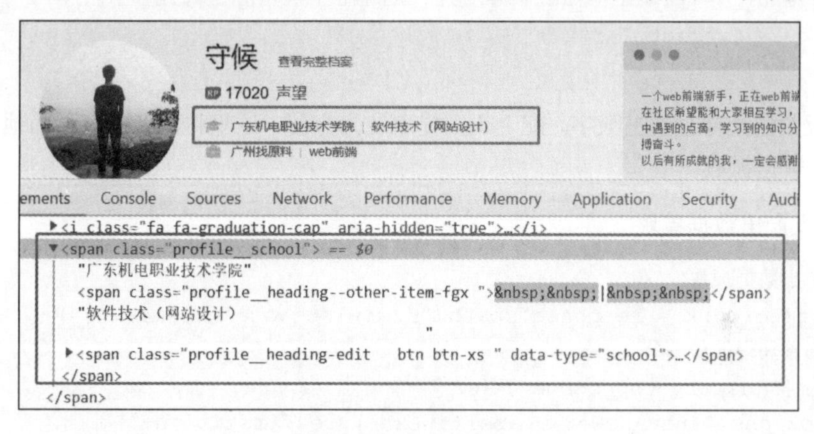

图 11.16 院校信息

学校专业在一个<span class="profile__school">标签中，使用 extract()提取标签中的文本时，文本列表第一项即为学校信息，第二项为专业信息。但需要注意一点，有的用户并没有填写学校信息，此项数据可能为空，在根据列表下标取数时就会报错。因此我们需要先做数据是否存在的判断，再来取数。如果为空，就填充空数据。公司数据也是一样处理的。

（2）个人统计面板数据

```
83 # 统计面板模块
84 profile_active = response.xpath("//div[@class='col-md-2']")
85 # 关注人数
86 item['following'] = profile_active.css('div[class*=info] a > .h5::text').
       re(r'\d+')[0]
87 # 粉丝人数
88 item['fans'] = profile_active.css('div[class*=info] a > .h5::text').
       re(r'\d+')[1]
```

```
89 # 回答问题数
90 item['answers'] = profile_active.css('a[href*=answer] .count::text').
          re_first(r'\d+')
91 # 提问数
92 item['questions'] = profile_active.css('a[href*=questions].
          count::text').re_first(r'\d+')
93 # 文章数
94 item['articles'] = profile_active.css('a[href*=articles] .count::text').
          re_first(r'\d+')
95 # 讲座数
96 item['lives'] = profile_active.css('a[href*=lives].count::text').
          re_first(r'\d+')
97 # 徽章数
98 item['badges'] = profile_active.css('a[href*=badges].count::text').
          re_first(r'\d+')
99 # 徽章详细页面地址
100 badge_url = profile_active.css('a[href*=badges]::attr(href)').
          extract_first()
101
```

统计面板数据大部分都是数字，配合正则提取较为方便，这里将徽章详细页面地址保存下来，后续进行徽章数据抓取。

（3）个人产出数据模块

```
102 # 产出数据模块
103 profile_work = response.xpath("//div[@class='col-md-7']")
104 # 回答获得的最高分
105 item['answers_top_score'] =
profile_work.css('#navAnswer .label::text').re_first(r'\d+')
106 # 最高分回答对应的问题的标题
107 item['answers_top_title'] = profile_work.css('#navAnswer
div[class*=title-warp] >   a::text').extract_first()
108 # 最高分回答对应的问题的 URL
109 answer_url = profile_work.css('#navAnswer div[class*=title-warp] >
a::attr(href)').extract_first()
110
```

同样，这里将问题的地址保存下来，后续进行详细的数据抓取。

（4）个人技能面板模块

```
111 # 技能面板模块
112 profile_skill = response.xpath("//div[@class='col-md-3']")
113 # 技能标签列表
114 item['skills'] = profile_skill.css('.tag::text').re(r'\w+')
115 # 获得的点赞数
```

```
116 item['like'] = profile_skill.css('.authlist').re_first(r'获得 (\d+)
        次点赞')
117 # 注册日期
118 item['register_date'] = profile_skill.css('.profile__skill--other
p::text').extract_first()
119
```

将注册日期放在此处一并抓取。由于注册日期有 3 种情况：注册于 2015 年 12 月 12 日、注册于 3 天前、注册于 3 小时前，因此不能简单地抓取数字，对于这种复杂的情况，需要在 Spider 中间件中进行处理，这个在后面进行介绍。

（5）二级页面抓取

最后，将个人主页没有抓取到的数据，如徽章详细数、回答的问题的详细内容，将要继续抓取的 URL 作为 meta 中的参数通过 Request 传递，并调用相应的处理方法，代码如下：

```
120 # 将需要继续跟进抓取数据的 URL 与 Item 作为参数传递给相应方法继续抓取数据
121 request = scrapy.Request(
122     # 问题详细页 URL
123     url=response.urljoin(answer_url),
124     meta={
125     # item 需要传递
126     'item':item,
127     # 徽章的 URL
128     'badge_url':response.urljoin(badge_url)},
129     # 调用 parse_ansser 继续处理
130     callback=self.parse_answer)
131 yield request
132
```

对于问题的详细页，我们需要抓取的数据有问题的标签、问题的内容以及最高分答案的内容，问题标签与问题内容定位如图 11.17 所示。

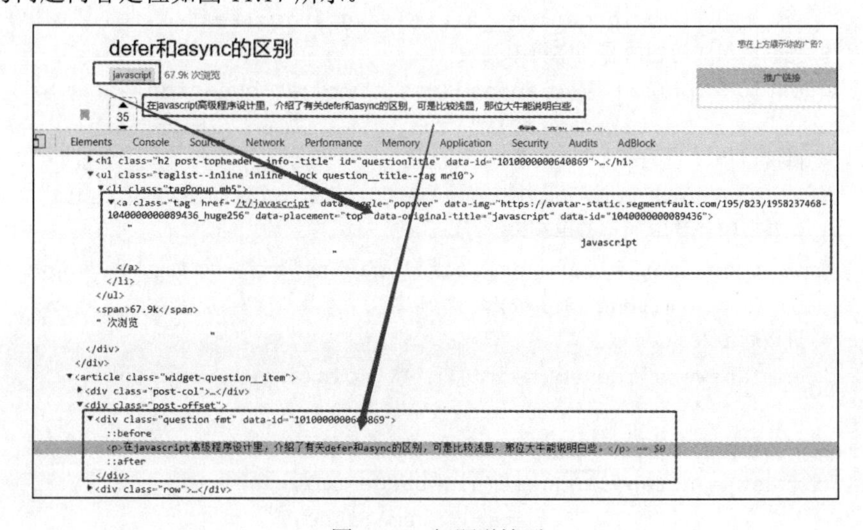

图 11.17　问题详情页

回答内容元素定位如图 11.18 所示。

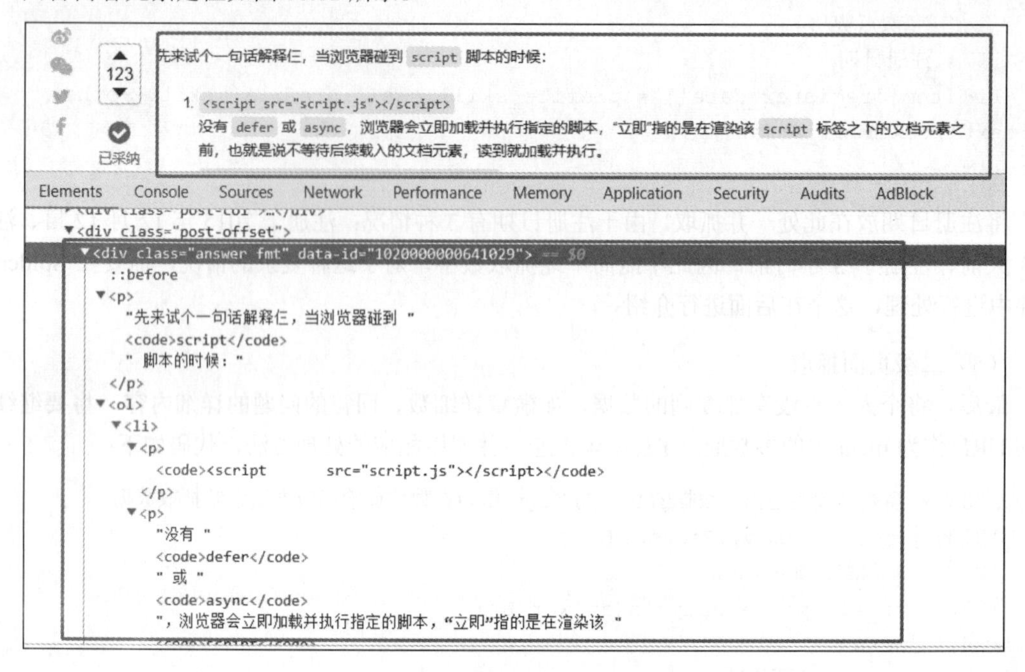

图 11.18 回答内容定位

userinfo.py 代码文件中 parse_answer()方法代码如下：

```
133 def parse_answer(self,response):
134     # 取出传递的 item
135     item = response.meta['item']
136     # 取出传递的徽章详细页 URL
137     badge_url = response.meta['badge_url']
138     # 问题标签列表
139     item['answers_top_tags'] = response.css
                  ('.question__title--tag .tag::text').re(r'\w+')
140     # 先获取组成问题内容的字符串列表
141     question_content = response.css('.widget-question__item p')
                  .re(r'>(.*?)<')
142     # 拼接后传入 item
143     item['answers_top_question'] = ''.join(question_content)
144     # 先获取组成答案的字符串列表
145     answer_content = response.css('.qa-answer > article .answer')
                  .re(r'>(.*?)<')
146     # 拼接后传入 item
147     item['answers_top_content'] = ''.join(answer_content)
148
149     # 问题页面内容抓取后继续抓取徽章页内容，并将更新后的 item 继续传递
150     request = scrapy.Request(url=badge_url,
151                        meta={'item':item},
```

```
152                          callback=self.parse_badge)
153     yield request
154
```

在抓取完问题页面的数据之后，继续传递 item 给 parse_badge()，抓取徽章数据进行填充，徽章页面需要抓取徽章名称与数量，如图 11.19 所示。

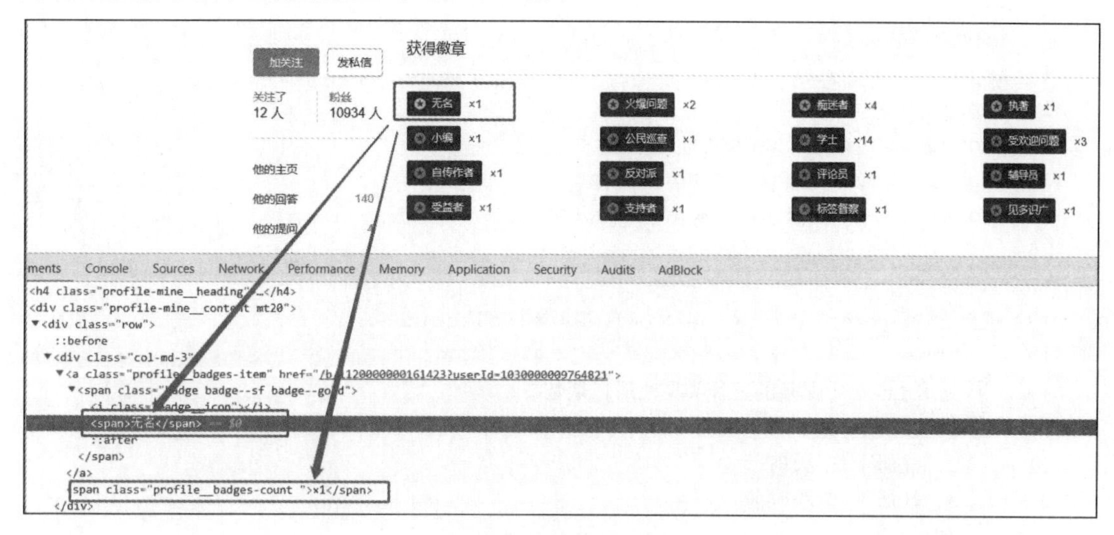

图 11.19 徽章详情页

parse_badge()方法代码如下：

```
155 def parse_badge(self,response):
156     item = response.meta['item']
157     badge_name = response.css('span.badge span::text').extract()
158     badge_count =
response.css('span[class*=badges-count]::text').re(r'\d+')
159     name_count = {}
160     for i in range(len(badge_count)):
161         name_count[badge_name[i]] = badge_count[i]
162     item['badges'] = name_count
163     yield item
```

将徽章名称与数量组成字典类型，将所有徽章统计做成列表存入 item 中，最后返回 item。

11.2.5 Middlewars 中间件编写

处理完爬虫文件之后，还需要进行中间件的处理。在前面的爬虫文件中，对于注册日期，因为数据的多样，无法直接提取数据保存，所以我们需要在 SpiderMiddlewares 中进行处理。middlewares.py 中 SegmentfaultSpideMiddlerare 代码如下：

```
01 # -*- coding: utf-8 -*-
02
```

```
03 # Define here the models for your spider middleware
04 #
05 # See documentation in:
06 # https://doc.scrapy.org/en/latest/topics/spider-middleware.html
07 import random
08 import re
09 import datetime
10 import scrapy
11 import logging
12 from scrapy.conf import settings
13 from pymongo import MongoClient
14 logger = logging.getLogger(__name__)
15
16
17 class SegmentfaultSpiderMiddleware(object):
18     """
19     处理Item中保存的三种类型注册日期数据:
20     1. 注册于 2015年12月12日
21     2. 注册于 3 天前
22     3. 注册于 5 小时前
23     """
24
25     def process_spider_output(self,response,result,spider):
26
27         """
28         输出response时调用此方法处理item中register_date
29         :param response:
30         :param result: 包含item
31         :param spider:
32         :return:处理过注册日期的item
33         """
34         for item in result:
35             # 判断获取的数据是否是scrapy.item类型
36             if isinstance(item,scrapy.Item):
37                 # 获取当前时间
38                 now = datetime.datetime.now()
39                 register_date = item['register_date']
40                 logger.info("获取注册日志格式为{}".format(register_date))
41                 # 提取注册日期字符串,如'注册于2015年12月12日' => '20151212'
42                 day = ''.join(re.findall(r'\d+',register_date))
43                 # 如果提取数字字符串长度大于4位,则为'注册于2015年12月12日'形式
44                 if len(day) > 4:
45                     date = day
46                 # 如果'时'在提取的字符串中,则为'注册于8小时前'形式
47                 elif '时' in register_date:
```

```
48                  d = now - datetime.timedelta(hours=int(day))
49                  date = d.strftime("%Y%m%d")
50              # 最后一种情况就是'注册于 3 天前'形式
51              else:
52                  d = now - datetime.timedelta(days=int(day))
53                  date = d.strftime("%Y%m%d")
54
55              # 更新 register_date 值
56              item['register_date'] = date
57          yield item
58
```

process_spider_output()方法在 Spider 输出 Response 时被调用，先判断 register_item 类型，再来使用 datatime 计算注册日期。now 为当前日期，使用 datetime.timedelta 将提取到的“3 天前”“5 小时前”，在参数中指定 days 或 hours，转化为 datetime 类型，再使用 now 减去该时间即为注册时间。最后使用 strftime()方法转化为“20171208”型字符串。

在 middlewares.py 中，我们还需要编写两个 DownloadMiddleware，分别处理 User-Agent 与 cookies。其中，SegmentfaultUserAgentMiddleware 在每次请求前向 Request 中随机添加 User-Agent。SegmentfaultCookiesMiddleware 有两个目的：一是每次请求添加 cookies，二是在发生重定向到登录页面时，替换原来的 cookies，并将原来的 cookies 删除。这两个 DownloadMiddleware 中间件代码如下：

```
59 class SegmentfaultUserAgentMiddleware(object):
60     def __init__(self):
61         self.useragent_list = settings['USER_AGENT_LIST']
62
63     def process_request(self,request,spider):
64         user_agent = random.choice(self.useragent_list)
65
66         # logger.info('使用的 USE USER-AGENT:{}'.format(user_agent))
67         request.headers['User-Agent'] = user_agent
68
69
70
71 class SegmentfaultCookiesMiddleware(object):
72     client = MongoClient(settings['MONGO_URI'])
73     db = client[settings['MONGO_DB']]
74     collection = db['cookies']
75
76     def get_cookies(self):
77         """
78         随机获取 cookies
79         :return:
80         """
```

```
81          cookies = random.choice([cookie for cookie in
                      self.collection.find()])
82          # 将不需要的"_id"与"_gat"参数删除
83          cookies.pop('_id')
84          cookies.pop('_gat')
85          # 将"Hm_lpvt_e23800c454aa573c0ccb16b52665ac26"填充当前时间
86          cookies['Hm_lpvt_e23800c454aa573c0ccb16b52665ac26'] =
                      str(int(time.time()))
87          return cookies
88
89      def remove_cookies(self,cookies):
90          """
91          删除已失效的 cookies
92          :param cookies:
93          :return:
94          """
95          # 随机获取 cookies 中的一对键值，返回结果是一个元组
96          i = cookies.popitem()
97          # 删除 cookies
98          try:
99              logger.info("删除 cookies{}".format(cookies))
100             self.collection.remove({i[0]:i[1]})
101         except Exception as e:
102             logger.info("No this cookies:{}".format(cookies))
103
104     def process_request(self,request,spider):
105         """
106         为每一个 request 添加一个 cookie
107         :param request:
108         :param spider:
109         :return:
110         """
111         cookies = self.get_cookies()
112         request.cookies = cookies
113
114     def process_response(self,request,response,spider):
115         """
116         对于登录失效的情况，可能会重定向到登录页面，这时添加新的 cookies 继续，
            将请 117 求放回调度器
118         :param request:
119         :param response:
120         :param spider:
121         :return:
122         """
123         if response.status in [301,302]:
```

```
124             logger.info("Redirect response:{}".format(response))
125             redirect_url = response.headers['location']
126             if b'/user/login' in redirect_url:
127                 logger.info("Cookies 失效")
128 
129                 # 请求失败，重新获取一个 Cookie，添加到 Request，并停止后续
130                 # 中间件处理此 Request，将此 Request 放入调度器
131                 new_cookie = self.get_cookies()
132                 logger.info("获取新 cookie:{}".format(new_cookie))
133                 # 删除旧 cookies
134                 self.remove_cookies(request.cookies)
135                 request.cookies = new_cookie
136             return request
137 
138         return response
```

在 settings.py 中添加 USER_AGENT_LIST 列表：

```
01 # User-Agent 列表
02 USER_AGENT_LIST = [
03     'Mozilla/5.0 (Windows NT 6.1; WOW64) AppleWebKit/537.36
           (KHTML, like Gecko)
04 Chrome/39.0.2171.95 Safari/537.36 OPR/26.0.1656.60',
05     'Opera/8.0 (Windows NT 5.1; U; en)',
06     'Mozilla/5.0 (Windows NT 5.1; U; en; rv:1.8.1) Gecko/20061208
           Firefox/2.0.0 Opera 9.50',
07     'Mozilla/4.0 (compatible; MSIE 6.0; Windows NT 5.1; en) Opera 9.50',
08     'Mozilla/5.0 (Windows NT 6.1; WOW64; rv:34.0) Gecko/20100101
           Firefox/34.0',
09     'Mozilla/5.0 (X11; U; Linux x86_64; zh-CN; rv:1.9.2.10) Gecko/20100922
           Ubuntu/10.10
10 (maverick) Firefox/3.6.10',
11     'Mozilla/5.0 (Windows NT 6.1; WOW64) AppleWebKit/537.36 (KHTML,
           like Gecko)
12 Chrome/39.0.2171.71 Safari/537.36',
13     'Mozilla/5.0 (X11; Linux x86_64) AppleWebKit/537.11 (KHTML, like Gecko)
14 Chrome/23.0.1271.64 Safari/537.11',
15     'Mozilla/5.0 (Windows; U; Windows NT 6.1; en-US) AppleWebKit/534.16
           (KHTML, like
16 Gecko) Chrome/10.0.648.133 Safari/534.16',
17     'Mozilla/5.0 (Windows NT 6.1; WOW64) AppleWebKit/537.1 (KHTML,
           like Gecko)
18 Chrome/21.0.1180.71 Safari/537.1 LBBROWSER',
19     'Mozilla/4.0 (compatible; MSIE 6.0; Windows NT 5.1; SV1; QQDownload
20 732; .NET4.0C; .NET4.0E; LBBROWSER)',
```

```
21    'Mozilla/4.0 (compatible; MSIE 6.0; Windows NT 5.1; SV1; QQDownload
22 732; .NET4.0C; .NET4.0E)',
23    'Mozilla/5.0 (Windows NT 5.1) AppleWebKit/535.11 (KHTML, like Gecko)
24 Chrome/17.0.963.84 Safari/535.11 SE 2.X MetaSr 1.0',
25    'Mozilla/4.0 (compatible; MSIE 7.0; Windows NT 5.1; Trident/4.0;
       SV1; QQDownload
26 732; .NET4.0C; .NET4.0E; SE 2.X MetaSr 1.0)',
27    'Mozilla/5.0 (Windows NT 6.1; WOW64) AppleWebKit/537.36 (KHTML,
          like Gecko)
28 Maxthon/4.4.3.4000 Chrome/30.0.1599.101 Safari/537.36',
29    'Mozilla/5.0 (Windows NT 6.1; WOW64) AppleWebKit/537.36 (KHTML,
          like Gecko)
30 Chrome/38.0.2125.122 UBrowser/4.0.3214.0 Safari/537.36'
31 ]
```

SegmentfaultUserAgentMiddleware 中的代码很容易理解，从 settings.py 配置的 USER_AGENT_LIST 随机获取一个 User-Agent 添加到 Request 中。

在 SegmentfaultCookiesMiddleware 中，我们添加了两个方法：get_cookies()与 remove_cookies()，分别用于从数据库中获取 cookies 与移除失效的 cookies。在 process_response()方法中，我们根据 response 状态码是否为 301、302 判断重定向，如果重定向到了登录页面，就说明 cookies 已经失效，重新添加 cookies，并删除原 cookies。

别忘了在 settings.py 中激活 SpiderMiddleware 与 DownloadMiddleware：

```
01 # Enable or disable spider middlewares
02 # See https://doc.scrapy.org/en/latest/topics/spider-middleware.html
03 SPIDER_MIDDLEWARES = {
04    'segmentfault.middlewares.SegmentfaultSpiderMiddleware': 543,
05 }
06
07 # Enable or disable downloader middlewares
08 # See https://doc.scrapy.org/en/latest/topics/
                                    downloader-middleware.html
09 DOWNLOADER_MIDDLEWARES = {
10    'segmentfault.middlewares.SegmentfaultUserAgentMiddleware':643,
11    'segmentfault.middlewares.SegmentfaultCookiesMiddleware':743,
12
13 }
11.2.6 run 启动器
```

前面我们都是从命令行启动爬虫的，有一个问题是不方便打断点调试。Scrapy 提供了命令行方法，利用此方法可以在代码中启动爬虫。在 setting.py 同级目录创建 run.py 文件，代码如下：

```
01 from scrapy import cmdline
02 from segmentfault.get_cookies import GetCookies
03
04 if __name__ == '__main__':
```

```
05        cookies = GetCookies()
06        cookies.save()
07        name = 'userinfo'
08        cmd = 'scrapy crawl {}'.format(name)
09        cmdline.execute(cmd.split())
```

Scrapy 中启动命令行的方法为 cmdline.execute()，将 cmd 命令传入执行即可。另外，我们添加 cookies 收集器 get_cookies.py，获取 cookies 执行爬虫命令即可。至此，整个项目代码编写结束。

执行 run.py，查看结果。保存的 cookies 如图 11.20 所示。收集到的用户数据如图 11.21 所示。

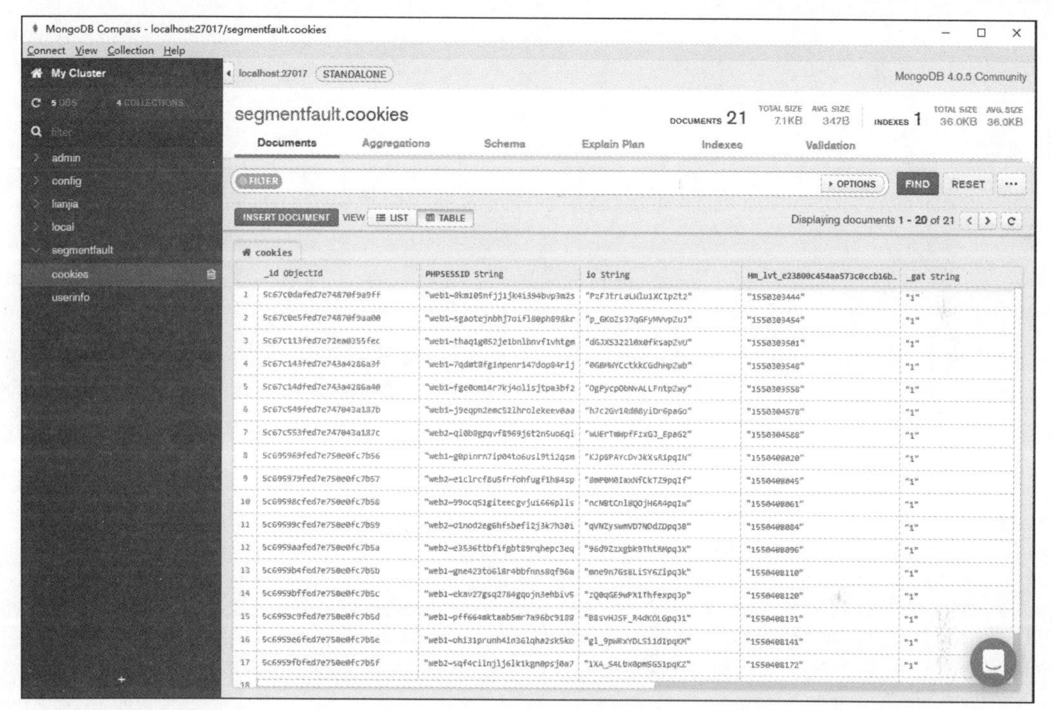

	_id ObjectId	PHPSESSID String	io String	Hm_lvt_e23800c454aa573c0ccb16b...	_gat String
1	5c67c0dafed7e74870f9a9ff	"web1~8km105nfjj1jk4i394bvp3m25"	"PzF3trLaLWlu1XClpZt2"	"1550303444"	"1"
2	5c67c0e5fed7e74870f9aa00	"web1~sgaotejnbhj7oifl80ph898kr"	"p_GKoZs37qGFyMVvpZuJ"	"1550303454"	"1"
3	5c67c113fed7e72ea0355fec	"web1~thaq1g052je1bnlbnvf1vhtgm"	"dGJXS322l0x0fksapZvU"	"1550303501"	"1"
4	5c67c143fed7e743a4286a3f	"web1~7qdmt8fg1npenr147dop84rij"	"0GBMWYCctkkCGdhHpZwb"	"1550303548"	"1"
5	5c67c14dfed7e743a4286a40	"web1~fge0om14r7kj4olisjtpa3bf2"	"OgPycpObNvALLFntpZwy"	"1550303558"	"1"
6	5c67c549fed7e747043a187b	"web1~j9eqpn2emc52lhrolekeev0aa"	"h7c2Gv1Rd0ByiDr6paGo"	"1550304578"	"1"
7	5c67c553fed7e747043a187c	"web2~ql0b9gpqvf8969j6t2n5uo6qi"	"WJErTmWpfFIxGJ_EpaG2"	"1550304588"	"1"
8	5c695969fed7e750e0fc7b56	"web1~g0pinrn7ip04to6usl9ti2gsm"	"KJp0PAYcDvJkXsRipqIN"	"1550408020"	"1"
9	5c695979fed7e750e0fc7b57	"web2~e1clrcf8u5frfohfugf1h84sp"	"8mP0M0IaxNfCkTZ9pqIf"	"1550408045"	"1"
10	5c69598cfed7e750e0fc7b58	"web2~99ocq51giteecgvjui666plls"	"ncN8tCnl8QOjH6R4pqIw"	"1550408061"	"1"
11	5c69599cfed7e750e0fc7b59	"web2~o1nod2eg6hfsbefi2j3k7h30i"	"qVNZyswmVD7NOdZDpq38"	"1550408084"	"1"
12	5c6959aafed7e750e0fc7b5a	"web2~e3536ttbf1fgbt89rqhepc3eq"	"96d9ZzxgbK9ThtRMpq3X"	"1550408096"	"1"
13	5c6959b4fed7e750e0fc7b5b	"web1~gne423to618r4bbfnns8qf96a"	"mne9n7Gs8LiSY6Zipq3k"	"1550408110"	"1"
14	5c6959bffed7e750e0fc7b5c	"web1~ekav27gsq2784gqojn3ehbiv5"	"zQ0qGE9wPX1Thfexpq3p"	"1550408120"	"1"
15	5c6959c9fed7e750e0fc7b5d	"web1~pff664mktaab5mr7a96bc9189"	"B8sVHJSF_R4dKOLGpq31"	"1550408131"	"1"
16	5c6959e6fed7e750e0fc7b5e	"web1~ohi31prunh4in36lqha2sk5ko"	"gl_9pWRxYDLS11dIpqKM"	"1550408141"	"1"
17	5c6959fbfed7e750e0fc7b5f	"web2~sqf4cilnjlj6lk1kgn0psj0a7"	"1XA_S4Lbx0pmSG51pqKZ"	"1550408172"	"1"

图 11.20　保存的 cookies

userinfo

	_id ObjectId	name String	rank String	school String	majors String
4	5c695dadfed7e710b4441284	"webfansplz"	"1303"	""	""
5	5c695db2fed7e710b4441285	"司徒正美"	"5024"	""	""
6	5c695db6fed7e710b4441286	"这是你的玩具车吗"	"2787"	""	""
7	5c695dbefed7e710b4441287	"沉静地闪光"	"4809"	""	""
8	5c695dbffed7e710b4441288	"zhoutk"	"2443"	"西安电子科技大学"	"计算机科学与技术"
9	5c695dc3fed7e710b4441289	"crossoverJie"	"3656"	""	""
10	5c695dc5fed7e710b444128a	"咚子"	"2840"	""	""
11	5c695dc6fed7e710b444128b	"dokelung"	"4824"	"國立台灣大學"	"電子設計自動化-應用邏輯
12	5c695dc6fed7e710b444128c	"数一数逍遥"	"415"	""	""
13	5c695dc6fed7e710b444128d	"东方小明"	"128"	""	""
14	5c695dc7fed7e710b444128e	"dreamans"	"539"	""	""
15	5c695dc7fed7e710b444128f	"廪冰"	"8291"	"厦门大学"	"软件工程"
16	5c778376fed7e751a42f79ca	"buildupchao"	"530"	"太原科技大学"	"软件工程"
17	5c778377fed7e751a42f79cb	"五月花开"	"1222"	"天津工业大学"	"光电信息科学与工程"
18	5c778377fed7e751a42f79cc	"LHK大帅仔仔"	"1545"	"广东工业大学"	"信息工程"
19	5c778380fed7e751a42f79cd	"webfansplz"	"1395"	""	""
20	5c778398fed7e751a42f79ce	"司徒正美"	"5032"	""	""

图 11.21　收集到的用户数据